Please return or
latest date l
MAY BE REN BY
D0077034

BRITISH TRADE UNIONS AND THE PROBLEM OF CHANGE

BRITISH TRADE UNIONS
and the Problem of Change

WILL PAYNTER
Secretary, National Union of Mineworkers
1959–1969

Montreal
McGILL–QUEEN'S UNIVERSITY PRESS
1970

HD
6664
P36

FIRST PUBLISHED IN 1970

This book is copyright under the Berne Convention. All rights reserved. Apart from any fair dealing for the purpose of private study, research, criticism or review, as permitted under the Copyright Act, 1956, no part of this publication may be reproduced, stored in a retrieval system, or transmitted, in any form or by any means, electronic, electrical, chemical, mechanical, optical, photocopying, recording or otherwise, without the prior permission of the copyright owner. Enquiries should be addressed to the Publishers.

© George Allen & Unwin Ltd 1970

ISBN 0 7735 0096 0

Library of Congress Catalog Card No. 75–123196

PRINTED IN GREAT BRITAIN
in 10 on 11 pt Times type
BY CLARKE DOBLE AND BRENDON LTD
PLYMOUTH

ACKNOWLEDGEMENT

My sincere thanks are offered to John Torode, Labour Correspondent of *The Guardian* for his invaluable criticisms and suggestions during the course of the preparation of this book.

W.P.

Contents

1

Introduction

THERE have been many studies of British trade unionism. This is another one, presented in the hope of provoking discussion on the strategy and structure of the British trade union movement in relation to its role and purpose in modern society.

It is hardly possible these days to open a newspaper or periodical without being confronted by some critical comment on one aspect or another of trade union activity or lack of it. The spotlight of stimulated public interest is concentrated directly and continuously upon them, and no opportunity is lost by the formidable band of critics to expose the weakness or misdemeanours, real or alleged, of its leaders and members.

Such criticism is, of course, mainly from sources hostile to trade unionism in general, and directed towards conditioning public opinion in favour of curbs and restrictions to restraining and limiting its freedoms. These people have a very definite axe to grind; the same political axe their kind have been grinding away at ever since trade unions were first formed. For them, criticism of a malicious kind is a form of self defence, a defence of class and privilege, of their right to property and profit, in short, a weapon used in the defence of capitalism, for the *status quo* in social relations, and against radical social change.

Although publicised criticism comes mainly from hostile sources, there naturally is criticism, too, from friendly sources, particularly from within the trade union and labour movement itself. This study is intended to be critical, although it it hoped that it will be seen as constructive and helpful. Into the study is brought such wisdom as one accumulates from a lifetime of experience in the trade union movement, and an awareness based on this experience of the deep-rooted opposition to 'change' that is to be found within unions. The old hymn sings of 'Change and decay in all around I see', but for many of us we form part of that which changes not, especially those personally treasured habits and interests around which our lives revolve.

This philosophy of 'changing not' is particularly strong in the trade unions concerning their structure and strategy. It is supported by the holy gospel as enunciated by many of the brethren in the trade union hierarchy – plenty of lip service to the desirability for change in both these matters coupled with a robust rearguard action, so that any change being effected is slow and ponderous. The mergers and amalgamations sometimes used by some trade union leaders, in an attempt to show that change in structure is in fact taking place, are really a rather pathetic apology for structural change, and an examination of them proves that the basic outline of horizontal unionism is being strengthened.

It is also a curious fact that union leaders of the 'left' appear as the diehards in this posture of opposition to changes which disturb the prevailing structure, or which produce a radical change in the role and policy of the movement. The assessment of economic and political change arising from my own experience over the years, convinces me that fundamental structural and strategic changes are urgently necessary if the trade union movement is to maximize its unity and effectiveness in present conditions. There have been decisive changes in the general environment in which unions operate, which call for a deep examination of these problems.

'Tradition weighs like an alp,' wrote Karl Marx, and it is certainly true of trade unionism. Who can honestly deny that there is with us a slavish adherence to the traditional; a habit of doing, accepting and believing, because this is what we have always done? In British unions, 'precedent' is worshipped as some kind of immutable law, and its absence, in relation to matters that arise, the source of serious dilemma. This is no exaggeration, as can be testified by those members of trade union staffs who have to look back over minutes and records searching for the elusive 'precedent'. The logic of this attitude is rooted, maybe not consciously but by habit, in the outlook maintaining that the answers to the problems of the present are to be found in the past, or that situations and problems that arise are similar to those that have gone before for which answers already exist. If this were the case, all a trade union leader would require is a 'ready reckoner' of 'precedents' and he could dispense with the need for original thinking.

Events, however, are compelling factors; those of recent date arising from the economic measures applied by Mr Wilson's Government, and arising from the Royal Commission on Trade Unions and Employers' Associations, are forcing continuous discussions, both within the Trade Union Congress and the individual unions, particularly on policy questions. The introduction of the Government's Prices and Incomes Policy, has caused the trade union movement, for the first time in any depth, to undertake its own analysis and to formulate for discussions with the Government its own proposals on the measures it considers necessary to stimulate national economic growth, to maintain full employment and financial stability.

As a result of the Donovan Commission's report and the White Paper and Bill now based upon it, proposing reforms in industrial relations in British industry, the unions and employers' organizations are having to review the whole maze of collective bargaining systems, and the wage structures they serve and with which they are associated. Any study that promotes discussion on these important matters is timely and useful if it helps to clarify policy issues and to bridge the deepening gulf within the movement.

If for no other reason, a discussion is necessary in the trade unions and the labour movement to dissipate the anger and condemnation directed against the Labour Government from the trade union side – by extending understanding of the nature of the problems with which the labour movement as well as the Government are confronted, and seeking a common policy to meet them that will unite and not further divide our ranks.

The situation is serious when wings of the movement, with ostensibly identical aims, are in deep conflict on issues of national economic and financial policy, and on the issue of the Government's intervention by legislation in industrial relations. The dangers of such conflict is most serious in its long term political consequences. It is not an accident that the Tory party seeks to take advantage of the differences that exist, and invites public support for legislation to put trade unions in a legal strait-jacket. This is a very real danger and prospect. The debate on the structure of the trade union movement and its strategy and tactics is highly relevant to a solution of this conflict and to the prevention of prolonged Tory rule, with the trade unions as its main target for attack on the economic front.

DEFINITION AND PURPOSE OF A TRADE UNION

'A trade union', said the Webbs in their *History of Trade Unions*, 'is a continuous association of wage earners for the purpose of maintaining or improving the conditions of their working lives.'

The term 'wage earners' is now too narrow a definition and must be interpreted so as to include employees of all kinds. But the Webbs state in simple terms the purpose for which unions exist. The pursuit of this purpose is inevitably influenced by the political and economic environment at any given time, as the history of the trade union movement in Britain provides ample testimony. It is not necessary for the purpose of this study to trace the origin and growth of unionism down through the years. It is sufficient to assess the changes in the environment affecting their operations over the last half century. This is convenient, too, because it covers the period of my own association with trade unions, most of it, from 1926 on, in some leading position or other.

The statement of purpose quoted above applies equally in a capitalist or socialist society. This study, however, is concerned with unionism under capitalist conditions in Britain.

It is true, of course, that over the past fifty years there have been many liberalizing reforms affecting employment in industry and relaxing the stringent restrictions on unions, easing the conditions within which they have functioned. But their purpose remains constant. A fundamental conflict of interest is inherent in the relation of capital and labour. It is, if nothing more, the characteristic market conflict within the capitalist economy as between buyer and seller. The buyer naturally wants to buy cheap and the seller to sell dear. This obviously does not by any means exhaust the conflict of interest that arises between worker and employer but it is a vital one, affecting their relations within industry.

Although the primary purpose of trade unions is to maintain and improve the working lives of their members, most unions include, in the objectives outlined in their constitutions, the political aim of 'abolishing capitalism'. This, too, is not just a pious or nebulous aspiration. The depth of feeling attached to this political aim was clearly demonstrated some years ago during the debate in the Labour movement on Clause 4 of the Labour Party constitution. This Clause calls for the realization of 'the social ownership of the means of production, distribution and exchange', modified somewhat now to taking over the 'commanding heights' of the economy.

Despite this, however, there is a school of thought within the Labour movement which considers that such great fundamental changes have already taken place in the character of the system that the term 'capitalism' is no longer appropriate. Mr Tony Crossland in his book *The Future of Socialism* apparently suggests this. But Mr Andrew Schonfield, the economist, in his book *Modern Capitalism* does not share this view and prefers to stick to the term because, as he states:

'I believe that our societies continue to possess many characteristics which are inextricably connected with their antecedents in the nineteenth and first half of the twentieth centuries; the word helps to emphasize the continuity. There are, after all, still large areas of economic activity open to private venture capital and in these areas its success or failure is determined by the familiar ingredients; the amount of liquid funds available, the efficiency by which they are manipulated, the personal initiative of the controllers of this private wealth and the enterprise of competing owners of private capital. Moreover, the prizes for individual success are still large and they convey on those who win them considerable economic power.'

Not only is the prize considerable economic power, but considerable political power as well. A fundamental change in capitalism involves measures to abolish private ownership of the means of production, distribution and exchange, and the political power associated with it. It is not therefore suggested that a review of union structure and policy is necessary because capitalism has changed. Capitalism is still capitalism and, although there have been changes, its basic character of exploitation and class struggle remains, although perhaps with reduced heat and less apparent hatred.

But acknowledging this fact does not detract from the proposition that changes have taken place which merit a review of policy by the trade union movement. I think the best way to illustrate these changes is to contrast the present with the nature of employer-union relationships during the inter-war years and the role of the state towards them. In doing this, I draw upon my own experiences and reminiscences of the period.

A PERSONAL TESTAMENT

I commenced employment in the coal mines of South Wales in the last month of 1917, upon reaching the age of fourteen. It was the practice in the coalfield for lads to go straight to the coal face as assistants to adult colliers, and this I did. I joined the local branch of the miners' union immediately, although I did not become more than casually interested until the 1926 general strike and lockout of the miners. Since I was working in the colliery next up the Rhondda Valley to the one where the branch leader was the stormy A. J. Cook – whose name was known even to boys as a fighter for the men – it was not possible to be unaware of the union and its local activities. During and following the lockout I became an active member and have remained so since. It may be argued that a lifetime in one union and one industry does not provide an adequate basis for a general view of unionism. The history of the mining industry over the last half century will show that it includes a wide range of the problems and vicissitudes of industry generally in very good measure, and more than enough to provide a rich background of misery and strife of all kinds, as the following brief account will show.

In the inter-war years the mining industry was, without doubt, the cockpit of industrial struggles in Britain and was in the forefront of the massive employers-trade union confrontations of the period. The period was one of recurring economic crises, with an almost continuous state of economic depression. Full employment was not a political aim of Government in those days. Unemployment, following the post-war reconstruction boom, ranged between 1922 and 1928 from 10 per cent to 14 per cent of the insured population, averaging through the period 1½ million annually. In the 1930s it was to soar to between 2 and 3 million, and was a period of widespread social dislocation and misery for millions of people, not only for those out of work, but for those in work as well.

Although the development towards greater concentration and centralization in economic organization was rapidly advancing, the dominant owners in industry were clearly associated with the day to day running and management of their industries. In mining, a South Wales coalowner, active in running his own pits, was chairman of both the South Wales and National Coalowners' Association, leading for them in all the discussions on wages and conditions for the whole of the inter-war years and as tough and ruthless as any employer anywhere, as the chonicle of events prove.

Management at production level was arrogant and arbitrary with trade union recognition generally tardy and limited. The situation between 1920 and 1930 can only be described as a state of continuous aggressive war by the employers against the workers. The miners took the brunt of the attacks.

In 1921, the employers in coal-mining served notice on all their workers that their contracts would terminate and that employment could only continue by agreement on new contracts involving substantial reductions in wages. Some years earlier, a Triple Alliance had been formed between the miners, railwaymen and transport workers to meet such a situation. They were to take action together if either was attacked. The decision to act together was taken, but, on the eve of the miners' lockout, the railwaymen and transport unions withdrew from the pact and the miners were left to fight alone. The lockout lasted three months and ended with a return to work and the acceptance of wage reductions.

In 1926, the miners were again selected as the first section of industrial workers to be attacked. They were regarded by the employers as the spearhead of resistance to any attacks and if they could be isolated and defeated, the employers could reasonably expect to defeat the rest. This time, however, the whole organized trade union movement was committed to the defence of the miners. The attack was anticipated and had been prepared for by both the employers and the Government for over a year, and it was known that the miners would again face the brunt of the attack.

The decision was for general strike to resist the attack, not just in defence of the miners, but because it followed that defeat for the miners would result in wage reductions for all industrial workers. This was the pattern being followed. The General Strike lasted nine days and was undoubtedly a magnificent act of solidarity, but it was called off by Congress leaders after nine days and again the miners were left to fight on alone. The issue was further reductions in wages and an increase by half an hour in the working day in the mines. The battle continued for some seven months, again ending in defeat with wage cuts and longer hours having to be accepted by the miners. The defeat of the miners was followed, as in 1921, by an assault against the wages of all industrial workers in the country.

The struggles waged by the unions during this period were defensive in character, aimed at holding on to the advances gained during the post-war reconstruction boom. It was a period of ruthless attack by the employers and desperate defensive action by the workers; a period when the whole atmosphere of industrial relations was permeated with bitter hostility and hatred.

Apart from two short periods of minority Labour government in 1924–5 and 1929–31, the inter-war years were dominated by Tory rule. Their role in the big struggles was characteristically that of the previous central government; seeking to appear in the role of mediator. Frequently they set up commissions of inquiry – there was almost an epidemic of them in the mining industry – into the causes of dispute and the conflicts in industry, or ministers made themselves available to act as arbitrators or mediators between the contending parties; in a word, generally adopting the role of an impartial, external institution which existed 'above the battle', concerned only with the restoration and maintenance of industrial peace.

The *Daily Herald* of July 30, 1926 carried this illuminating report of an interview between miners' representatives and the Prime Minister, Mr Stanley Baldwin.

Miners: 'But what you propose means a reduction in wages.'
Prime Minister: 'Yes, all workers in this country have got to face reduction of wages.'
Miners: 'What do you mean?'
Prime Minister: 'I mean all workers in this country have got to take reductions in wages to help put industry on its feet.'

This method of 'putting industry on its feet' was not original then, and is still used in one form or another. So, behind the façade of impartiality, the Government supported the employers. In fact, their proposals for a settlement of the dispute in the mining industry during 1926 involved acceptance of big wage cuts and longer hours. Besides, in June of that year the Lord Chamberlain, Birkenhead, put forward plans to a Conservative Party meeting for restricting the rights and freedom of the unions. These proposals a year later were embodied in the Trades Disputes and Trades Unions Act, 1927. This Act outlawed general strikes, sympathetic strikes, or any strikes that could be construed as likely to coerce the Government directly or indirectly. The Act also prohibited picketing and sought to hamstring the Labour Party financially by amending the law affecting union political funds. Previously, political contributions could be collected from the trade union dues of all members except those who signed forms to 'contract out'. The new law laid it down that such political contributions would only be made for members who 'contracted in'. In other words, the member had to sign an authorization form before the payment could be made. This was a much more difficult position, as those of us who stood at the pit heads when the Act became law and had to get members to sign for the political levy payment were not long in finding out.

The climax of this whole period culminated in the 1931 'economy cuts' which imposed a 10 per cent reduction in the wages and salaries of all government employees, including the police, the army and the navy, and even the benefits of those unemployed. The minority Labour Government disintegrated, with Macdonald and Snowden and others forming a National Government with the Tories; the sailors in Invergordon mutinied; and a stage of general agitation affected the whole nation. This climax was a natural sequence of the collapse of resistance to the attacks on wages and conditions throughout the previous decade and the withdrawals and retreats of 1921 and 1926. Nevertheless, the involvement of the State in industrial affairs was in the main confined to imminent or actual industrial conflict. The initiative in incomes policy (to express the situation then in modern terms) was with the employers, the State acting in a supporting capacity. The factor of heavy unemployment obviously facilitated the policies being pursued by both the employers and the State.

The decade immediately before the Second World War was little different from the one that preceded it. There were strikes in many industries, but not of the dramatic national character of the earlier period. Many of them were for trade union recognition, especially in industries where unions had not previously been established. The combination of mass unemployment and the lowered morale consequent upon the defeats suffered in the 1920s had an enfeebling effect on unionism from which it was slow to recover.

The process of attacking wages was by no means exhausted. We had a short bitter struggle early in 1931 in the South Wales Coalfield against a further cut in wages. The dispute had been referred to an arbitrator who awarded in favour of the owners, recommending further reductions in wages and the introduction again of the subsistence principle. In this case a married man, doing the same job as a single man, was paid 3d a shift or 1s 6d a week extra!

At this time I had been elected to the position of checkweighman at the colliery where I worked. We were on strike for three weeks in the January of 1931 when a coalfield conference, on the advice of the leaders, decided to recommend acceptance of the reductions. There was an uproar in the Rhondda union branches, most of whom decided to continue the strike. At a mass meeting of the men at my pit, despite the fight we put up, the decision, or more accurately, the ruling of the chairman on the voting, was that we accept the conference recommendation to end the strike. After hearing the decisions of the other Rhondda pits, I proceeded on the following morning to hold a pit gate meeting to get solidarity action with the pits continuing the strike. I was forcibly ejected from the colliery premises, and some weeks later the colliery company obtained a court injunction, under the Checkweighman's Act of 1860, to have me removed from my post. I spent the next years in unemployment and there were more of us in that state in the Rhondda than there were in employment. The story is only significant now because it throws some light on the character of the relationships then and the reprisals taken by employers against union men.

In retrospect, the struggles that took place in industry during the 1930s were overshadowed by the movement against unemployment and especially the Means Test. This latter device was introduced with the economy cuts in 1931 and was directed against those in employment from families which included some who were unemployed. Under the test the total income of the family was assessed against its needs, as measured by the Government's predetermined scales, and the benefit paid to the unemployed member of the family was determined by this method. It meant, in effect, that part of the income of the employed was used to maintain the unemployed. It caused, as can be imagined, a great deal of internal friction within families and produced devious methods for deceiving the investigators.

It is perhaps of interest to recall the controversy within the Labour movement on the issue as to whether Labour activists should take jobs as Means Test investigators. Those in favour of doing this, and their viewpoint prevailed, argued that by taking up these appointments the means test would be operated 'more humanely'. This undoubtedly had the effect of disarming resistance, with the result that the resistance movement was slow to grow and make impact.

The highlights of the struggles of the unemployed were their several hunger marches. The initiative in their organization was with the National Unemployed Workers' Movement. The Trades Union Congress in 1931 was held in the Colston Hall, Bristol, and I took part in a march from South Wales to Bristol, as an unemployed man, urging the trade unions to take more positive action against the means test for work or full maintenance for the unemployed. We had a pretty rough reception both by the police and the Trades Union Congress.

We had marched from Newport to Bristol, riding only through the Severn Tunnel, and were very tired when we arrived for a meeting with the Bristol unemployed at the 'Horse Fair' in the centre of the town. When the meeting ended we formed up to march to a chapel vestry where we were billeted for the night. The police intervened to say we could not march, but after some discussion agreed that unless we went together to the billet many of the marchers might get lost, so we were allowed to march, but without banners or bands. Immediately we got to the main street outside the meeting place we were set upon by the police, who obviously were acting on original instructions and had not been advised that they had been changed. There was a free fight but I did not see much of it due to being carried into the police station. I was charged and fined later with 'attempting to rescue'. The Bristol press the following day carried bitter attacks against the police for the way they had acted and their indiscriminate attacks on non-marchers too.

The deputation that tried to get into the Congress Hall to put the case was ejected, and Wall Hannington, the unemployed leader, was thrown down the steps leading into the hall. Marches became almost an annual event between 1930 and 1936. In November 1932, we marched to London to lobby Parliament on the same issues. The reception we had when we arrived in London was tremendous.

On a particular Saturday, the contingent of marchers from the various parts of the country assembled in Hyde Park for a rally. I remember the thrill we had arriving at Marble Arch from Shepherd's Bush and marching in full kit through a lane of cheering men and women who were held back by legions of policemen with arms linked to form a barrier on both sides through which we could march. I was the treasurer for the Welsh contingent responsible for the collecting boxes. While the meeting was in progress they were left in a lorry in the care of the driver. When the meeting ended I went to collect them to continue the collecting on the march back, to find them in the process of being taken away by the police. The driver had wandered off and left them and he had apparently broken some by-law by parking the lorry inside the park. I asked the police what they were up to as I was responsible for the boxes, and they dropped the boxes and made a grab for me, and I made a dash for freedom. There were too many police about and I was caught and taken to the police station in the middle of the park. The inspector of police in charge was a Welshman, to whom I gave the marchers' money in my possession. I was bailed out sometime in the middle of the night, the marchers' leaders not knowing what had happened to me, and after I had been regaled by a drunken prostitute in the next cell loudly proclaiming the diseases from which royalty had suffered through the ages. I was brought up before Marlborough Street Magistrate's Court the following morning, charged with obstructing the police, fined £5 or a month's imprisonment, refused time to pay, being told by the Bench that I was a vagrant with no visible means of support, but granted time to consult the marchers' council. They had to decide whether they would pay the £5 or let me go down. Both points of view were expressed, but they finally decided to pay, mainly because the lorry and boxes had been confiscated by the police and were in my name and could not be recovered until I was free.

The interesting part of this story is that when I went to the police station in Hyde Park to retrieve the boxes and lorry, it was to discover that the Home Office was considering 'more serious' charges, including unlawful assembly and that our property was being held until a decision was reached. It was even alleged that our lorry was full of lethal weapons like pick shafts with 6 inch nails through them. This, of course, was a complete fabrication and could not have been sustained unless such weapons had been planted. There were no charges preferred and we finally had our property returned. The sequel to this attempt at prosecution of the marchers' leaders appeared some weeks later in a column in the *Reynolds News* (a Sunday paper of the Co-operative Party with progressive views) called 'Secret History', which claimed that the Welsh inspector of police had been prematurely retired for his sympathies with the Welsh marchers. Whether this was authentic or not is not possible to prove. The other little feature was that my arrest produced a banner headline in a South Wales evening paper, purporting that a Welsh unemployed marcher had been arrested with over £40 in his possession! The money, of course, was the collections made *en route* to buy food for those on the march.

The biggest and most spectacular march took place in 1936 and was supported by an effective united front of working class organizations. We were over five hundred marching from South Wales and our marchers' council included Aneurin

Bevan and other Members of Parliament. We were in London on November 11th when the ceremony of laying the wreath to the memory of the fallen in the First World War took place. By some machinations we were able to march past the Cenotaph behind the colourful Guards' Bands, investing the pageantry of the occasion with an ironic facet of the world that was supposed to be fit for heroes to live in. Nevertheless, it was a thrilling experience only surpassed in my experience by demonstrations in the Rhondda in 1935 and 1936 where we had built up a large measure of unity of industrial and political organizations. Sunday demonstrations against the Means Test were supported by many of the chapels and churches, who closed their Sunday schools to take part in the demonstrations. During those occasions the whole population of the Rhondda was on the streets and it was possible on these Sunday afternoons to stand on a field near the centre of Tonypandy among thousands of demonstrators who had already arrived and see, still coming over the mountain from the little Rhondda four miles away, an apparently endless river of people. This was the mass hatred that was generated against the Means Test applied to the unemployed.

In the build-up of action as far as this stage, however, there was quite a lot of police repression and many of the unemployed leaders suffered imprisonment and police persecution. But out of it there grew massive united front and popular front movements, in the vanguard of which was always to be found the redoubtable Aneurin Bevan, then a MP sponsored by the South Wales miners, who, on the many occasions that he was carpeted by Transport House, was assured of the support and protection of the Welsh miners.

During all this turmoil and conflict, the organized trade union movement was slowly recovering from the battering it had taken in the stormy 1920s. In 1920 its membership stood at 8·3 million; By 1922 it had fallen to 5·6 million and was to fall still further to 4·8 million by 1930; by 1938 it was rebuilding and reached the 6 million mark, but it was not until 1943 that it again topped 8 millions.

As to its status and standing with the central governments

of the day, the answer is best given from the Centenary Brochure of the TUC under the section described as the 'TUC's formative years and the fight for the right to be heard'.

'But on the major issues of Trade and Industry, poverty and and unemployment, collective security and rearmament, peace and war, the TUC during the 1930s found it almost impossible to get a serious hearing from the National Government, the Baldwin Government or the Chamberlain Government.
'Ever since the General Strike, Conservative governments had instinctively treated representatives of organized labour as men on the outside looking in.'

In fact, even after some limited form of consultation had started in the early days of the Second World War, when the unions asked for an assurance that the 1927 Trades Unions Act would be amended, they were told by Chamberlain that they were still on probation. It was only after Mr Ernest Bevin and other labour leaders had joined the Churchill War Government that permanent machinery for consultation between the TUC and the Government really got under way.

In summing up the characteristics of this period, as a basis for contrast with the present, its salient features can be put briefly and concisely.

Economic organization in industry, although in the process of concentration and centralization, had not reached giant size. Ownership of industry was identified with management and was directly involved in its day to day running. Management in practice was arrogant and arbitrary in its dealings with workers and unions and the battle for recognition and trade union freedom was still being fought.

Industrial relations, as typified by the processes of collective bargaining were non-existent, non-functioning, or an arena for bitter exchanges created as a by-product of the employers' assault on wages.

Central government was antagonistic to the trade union movement, relegating it to the wings as far as consultation and participation in the determination of policy was concerned. There was no attempt at economic planning or achieving full employment, and living standards were the subject of repeated attack. Wages and conditions of employment were a direct issue for settlement between employers and unions, with the Government indirectly supporting the employers.

The trade union movement was relatively weak in membership and influence, without adequate, well-equipped central leadership.

The foregoing is a brief description of the situation in the inter-war years, coloured somewhat, perhaps, by personal experience which at least should be as authoritative as a description based on an academic study. It is hoped that it provides sufficient of a background against which to assess and measure a comparison with the present.

2

Changes in Economic Structure

DURING the post-war years there has been an enormous growth in concentration and centralization of economic organization. Although there is still a multitude of private enterprise undertakings, varying in size from the very small to the very large, the emphasis is rapidly moving more and more to the very large economic organization in both industry and commerce. The growth of big economic organization does not necessarily mean a change in their social content. In the main, they remain essentially private enterprises, firmly based on private property ownership, with competition as the operating method of distribution, and their main motivation still dynamically that of rent, interest and profit.

The traditional relationship between them as employers and the trade unions is fundamentally unchanged, although there has been a significant shift in the balance of power between them. In the same way the aims of the unions remain unaltered, but with the changes in organization in industry and commerce and the power they represent becoming a growing challenge to the efficiency of trade union structure in industry, as well as to the effectiveness of their strategy. The growth of giant economic organizations, operating vertically and horizontally, overlapping each other in large parts of the economy of the country, at least poses for the unions the question as to whether or not their structure, strategy and tactics, born and fashioned in the past, are not, in relation to these new power centres, outmoded and inadequate.

Without doubt the bigest development towards giant organization in the post-war period has been the establishment of public corporations in a number of industries. The Labour Government following the war embarked on a programme of nationalization. The industries of coal, electricity generating and gas were taken over into public ownership; in more recent years atomic energy production and the exploitation of natural gas in the North Sea are, to a large measure, under the control of the Government. Thus the main source of energy supplies for the nation are publicly controlled. In transport the rail, air and road services are roughly in the same position. More recently, steel has been re-nationalized, making perhaps the most important and strategic penetration into the holy ground of private enterprise. This development extends the boundaries of public ownership far beyond those previously obtaining, and represents a big inroad into pure capitalist domination of the economy.

The combination of the activities of public corporations, together with those of central and local government, now represents a substantial proportion of the total wealth production of the nation. The TUC Economic Report of 1968 stated that over 6 million people, almost one quarter of the insured population, were employed in the public sector. Although there is no universally adopted method of assessing public expenditure, the official handbook *National Income and Expenditure* gives a crude figure of the proportion of public expenditure in the Gross National Product. Of the total GNP, i.e. national wealth production of over £34 thousand million, around £17 thousand million approximately was accounted for by the current and capital accounts of the Public Sector. This is roughly 46 per cent and obviously represents a formidable influence upon the economic and financial policies within industry and commerce. The Public Sector in this context includes the expenditure by central and local government on defence, social services, housing, education and so on, as well as that of public corporations.

The importance of the extension of public ownership is not that it represents socialism, as is sometimes claimed in windy political perorations. A socialist island within a national sea of capitalism is an impossible combination, taking the doctrine of coexistence into the realm of absurdity. This is not a sane judgement of the advantages of public ownership as a progressive measure. In fact, to those who agitated for this reform over the years, it turned out to be something less than our vision promised. For in our visions we saw it as a crippling attack on the citadels of capitalism; as the herald of its speedy demise and stepping stone to the new social order we dreamed about.

I remember the morning of January 1, 1947 – the first day of operations with nationalized mines – standing on the top of a tram of coal at one of the collieries I served as a union agent, making an enthusiastic speech about 'the dawn of a new era', of the significance of the day being one where the 'workers were moving forward to the control of their own destinies' and that we were at the beginning of the process where capitalism would be replaced by Socialism. This was obviously a naïve and immature judgement of the change taking place, as we soon realized from experience. The relations between management and worker remained the same; the union still had to fight hard to get improvements in wages and conditions and little in the daily lives of the men reflected the change that had taken place.

It would be wrong, however, to give the impression that everything remained unchanged. The change was mainly in the climate between management and the unions. Discussions within the conciliation machinery became less acrimonious; machinery for consultation provided an opportunity for workmen to influence management policy in a way that did not exist before. In the beginning, the improvements in industrial relations were more abstract than real and could not be measured by practical results; but as time went on and confidence grew, the union became a force that had to be consulted on all questions of major policy, becoming indirectly involved in the industry's management.

Nevertheless, although nationalized, these industries were a subordinate part of the total economy, expected to operate as commercial undertakings, generally on strictly business principles characteristic of capitalism. In the case of the mining industry, it became the milch cow for the privately owned sector, supplying fuel at less than world prices and often below cost in response to Government-directed economic policy. That was the first decade of coal-mining nationalization; in the second decade it was made the whipping boy for the economic ills of the nation. However, the experience in the mining industry is not a just and adequate basis for judging the advantages of nationalization generally.

In making the comparison between the present situation and that which obtained between the two world wars in the extension of public ownership, the change is indeed substantial. It means that a large part of the economy is no longer under direct capitalist control, and is no longer the source of profits and dividends for the big or small investors.

MERGERS AND TAKEOVERS

Second to the extension of nationalization is the rapid growth of the concentration of capital in big undertakings. The main medium of this development is through mergers and takeovers, a development that is actively encouraged by Government policy. The Commission for Industrial Reorganization has been established to stimulate this process toward bigger economic organization based upon bigger concentrations of capital. The purpose is to streamline industry and commerce based on big units employing the most up to date techniques, pruning the uneconomic and inefficient and generally enhancing the competitive ability of the particular goods being produced. In terms of capitalist economics the purpose is to concentrate capital investment as the means to higher profits; to buy out competitors and to cut costs through concentration; to facilitate expansion into new markets and to protect and defend the markets already held. The extent and rate of growth of this development is seen in the colossal sums involved in takeover bids. The estimated sums involved for 1966 is put at about £500 million; for 1967 the estimate is doubled to £1,000 million; for 1968 one estimate given is around £4,000 million, four times that of the previous year. These are estimates based on the figures that hit the headlines, but is probably an underestimate of the actual position.

These takeovers are not confined to any one industry but are spread pretty widely over all sections of the economy, certainly the sectors of importance. A list of takeovers in the last few years would include a variety of operations, from computer production and electronics to newspapers, as well as food, drugs, chemicals, shipbuilding, motors and their accessories, building and construction and engineering. The effect of these is to create powerful industrial and commercial giants in all the vital sectors of the economy. These great power centres spread throughout the economy, strengthen monopoly control and are able to radiate policy and influence over standards of production, costs and prices for the sectors of the economy in which they operate.

If this is a fair estimate of the changes taking place and their effects, which must increase as the development proceeds, then the question posed is whether they represent a new concentration of power by the employers that substantially changes the environment in which trade unions function and the balance of power between them. In my judgement they do and require similar concentrations of power on the trade union side.

There is an apparent contradiction in Government policy towards mergers and takeovers. The Commission for Industrial Reorganization exists to encourage mergers by offering loans and other inducements, while the Monopoly Commission exists to investigate developments towards monopoly, with its price-fixing and other antisocial propensities. In fact, during 1967 the Board of Trade asked that no less than ninety mergers and takeovers be referred to the Monopoly Commission, obviously out of fear of their anti-social potential, but only one was actually dealt with by the Commission and that was not a very important one. This was a year, too, when colossal sums like £900 million in one takeover deal were involved.

Ethics have never been a very strong feature in any of the operations and activities of capitalists and this is sadly true of those involved in takeovers. The 'poor victims' of the bid seem to attract a certain sympathy, but it is usually a highly profitable experience of 'victimization'. Shares can double their value in a matter of days and if there is more than one contender for the spoils they can be played off against each other to some advantage. The capital gains for the victims so called are generally rich pickings and the power and control gained by the bidder is expected in the long term to be worth the price paid. It may be considered a necessary development in economic organization, but it is certainly a pretty shoddy business, being about the moral equivalent of the dealings between the thief and the 'fence'. It is an area of financial manipulation in need of control and if bigger economic organization is required, it should be according to State planning and control.

It is, however, a fact of life and one that looks like continuing for some time. The new giants are not within one industry but overlap into others.

Mr John Hughes, lecturer in trade union and economic subjects at Ruskin College, wrote as far back as 1964 in *Change in the Unions*:

'By the end of 1960 "giant" firms with assets of £25 million or more accounted for the following proportion of total assets of over £0·5 million; Chemical and Allied 84 per cent; metal manufacture 74 per cent; retail distributors 66 per cent; vehicles 64 per cent; and electrical engineering 59 per cent.'

If this was the position in 1960 it does not require a very vivid imagination to appreciate what the position must be like now. As the estimates already referred to show, over the last three years the increase in takeovers and mergers has multiplied annually. The conclusion drawn by John Hughes four years ago must be much more true of the present when he wrote:

'These new industrial giants are not themselves tidily confined within recognizable industrial frontiers . . . and this produces new stresses and strains in union organization and estabished procedure at bargaining.'

It is this increase in 'stresses and strains' affecting union organization and functioning, arising from this factor of change that merits the study of union structure and policy.

PROFESSIONALISM IN MANAGEMENT

With this new growth of massive organization the identity of ownership has become shrouded in a maze of financial intricacy. The larger the undertaking the greater the involvement of banks, finance houses and investors of all kinds. Ownership has thus become 'depersonalized' and remote and is rarely seen to be involved in the actual direction and management of these big organizations. Compared with the early years of capitalism, ownership has been divorced from the day to day running of industry and is consequently more parasitic than ever! As a factor contributing something to actual economic operations it is obsolete, outmoded and completely unnecessary. The whole trend in industrial and commercial management is towards professionalism. Management in economic and social undertakings has become an important and highly specialized responsibility, requiring special techniques and special training. These managers are the steersmen of industry, the brains responsible for the day to day operations.

Ownership as such has become so diffused, capital investment so wide in source, the money of the big investors so diversified in its investment, that ownership has become nebulous and abstract. This is true at least in the context of industry. It follows therefore that the pressures that were effective against owners or employers in the past are not necessarily effective now. What might have been effective fifty years ago against an employer could today be no more than an irritant to the dominant shareholders who will have investments spread over other undertakings. The action or pressure is likely to be more damaging to the economy as a whole, with a boomerang effect upon the working people. This is not intended as an argument against trade union action and pressure; rather it is intended to point out that action in modern conditions needs to be studied and applied with care and after thoughtful preparation.

The other important feature of this change is the growth of foreign investment, especially American, in the British economy. The counterpart to this, of course, is British investment in other countries. With this spread of investment into other countries and from other countries into this one, a situation arises where action to bring one unit or a series of units to a standstill does not necessarily seriously impede the operations of the particular giant organization. It can cause dislocation in a particular country and while profitability of such an international organization might be affected, it is not necessarily brought to a halt.

An example of this can be taken from the motor car industry in this country. A large part of it is controlled by American capital, by companies with motor car production industries in a number of countries. Bringing car production to a standstill in one of these countries, whether it is by industrial action in the major plant or in the subsidiary plants producing the necessary accessories, does not immediately or necessarily stop car production by this company. It can, however, cause serious financial problems for the government of the country concerned, especially where motor cars are a main export. This aspect of the new situation has helped to produce the position where central government has moved into the arena of industrial relations. Many other examples could be quoted.

This kind of employers' organization spreads its economic organization not only over industries in one particular country, but in other countries as well. Their profit earning tentacles suck lucrative proceeds from many different economic arteries, and because of this, their resilience in face of attack is enormous, while their vulnerability to sporadic, sectional attack is small. The point being made is that the term 'employer' does not have the same connotation now as it had, say, fifty years ago. The original interpretation in the context of industry was that of identifiable persons. A modern interpretation in practice, at plant level, brings it down to the 'boss' who is usually the manager of the concern, or the more elevated managing director, both of whom may be professionals employed by the giant company. The situation has radically changed and the by-product of the change is the involvement of the State in industrial affairs far more directly than at any time previously.

THE TECHNOLOGICAL REVOLUTION

The changes in the form and size of economic organization synchronize with the changes in technology. Indeed, these changes in the type of the economy make possible the application of the new techniques to the production process. The potential involved in this technological revolution is tremendous. The most revolutionary element in this change is 'automation'.

I remember trying to give some idea of the changes inherent in the application of automation to the Miners' Annual Conference in 1963. I had been reading the papers read to a conference held in Geneva in December, 1962, promoted by the International Labour Office, where the papers were presented by spokesmen for the 'American Foundation on Automation and Employment'. This foundation was started and financed by the late John Snyder, a big industrialist and the head of a firm producing automated equipment. Later he helped to set up a similar foundation in this country for the purpose of studying the effects of automation upon employment. He was apparently a tycoon with a conscience.

In the paper he read to the Geneva conference he debunked the theories popularly spread in the United States to soothe anxieties about the effect of automation:

'The first myth I would like to talk to you about concerns the employment situation. Recently in America anyone who says that automation is going to put people out of work has been considered a prophet of doom. Some experts say that automation actually creates jobs; that everyone will always be employed because it takes people to make machines and keep them running. Well, I disagree. . . .' 'that might have been the case in the long run, but as Lord Keynes suggests, in the long run we are all dead. Today, Detroit is one of the centres of automation and it is also one of our country's largest and most critical areas of unemployment. In Detroit at least automation has not created jobs.'

At the same conference a speaker for the American unions reported that the attitude of the unions was hardening and they were insisting that this technological advance must be controlled and regulated and that necessary social change should synchronize with it. He reported that the unions were pressing for a 35 hour week as a means of accommodating displaced labour into other jobs.

While it seems that the effect on employment has not been as great as Snyder forecast, or as was the experience in Detroit, it has to be acknowledged that the essence of the new techniques based upon automation means the increase of global production with less labour required for each unit of production. The reduction in the labour force for the same output in the vast majority of situations where automation or semi-automation is introduced is, of course, influenced by the character of each operation. Generally speaking, however, the reduction in the labour required is substantial. There would be little point in making the capital outlay, usually very heavy, if there was no saving to be made in some other aspect of the operation and this is usually applied to labour costs. The effect in redundancies can be substantial, unless automation is accompanied by an expansion of operations as a whole, creating new job opportunities.

The best presentation I have found of the ultimate potential of automation is in the preface to *Modern Automation* by Dr John Foster, a book he wrote in 1963 after studying the development of automation in America, Russia and Europe. He defines automation as the practical aspect of 'cybernetics'. He writes:

'Now cybernetics means the art of steersman, but the technical implication is that, in the future, robot electronic brains will direct (or steer) all physical industrial processes towards perfection, whilst mere humans will concentrate on solving the problems of excess leisure.'

'Mere humans' will have to take steps to ensure that the 'excess leisure' is not in the form of mass unemployment, which is a certain outcome in an uncontrolled economic system geared to the pursuit of private profit. What is really forecast is the application of electronic brains, of which the computer is an example, to run the machines of industry. This may seem a far-fetched piece of romantic dreaming but there are plenty of instances now where this kind of control system operates.

I can best draw examples from the industry I know best, an industry that is certainly not the easiest in which to apply advanced techniques because of the absence of stability and continuity of physical environment, due to geological conditions. Nevertheless, the changes in technology applied now at the coal face and outbye, can only be described as dramatic. The completely manless operations envisaged in the application of 'robot electronic brains' has not been achieved yet, but there is every expectation that – with the mastery or the development of a device to control the advance of machines from one end of the coal face to the other, adjusting the machine to meet undulations that occur in the floor and roof of the seam – this manless face operation will be accomplished. This obviously does not mean that no men will be required at all; the maintenance and the monitoring of the equipment will still require trained men, but the main element of physical labour will be eliminated.

In the mining industry semi-automation is now widespread and is being extended to a whole variety of mining operations. For one who started in the pits over fifty years ago, as I did, when every particle of coal had to be cut and loaded by physical labour, when it was the muscle, sinew and sweat of the miner alone that was the applied power to production, the changes are really revolutionary. I have sat at the entrance to a coal face two hundred yards long with a seam height of three feet and looked down the face, equipped from end to end with lights, with its cutting and loading machine mounted on the conveyor structure which transports the coal to the loading point, to see the conveyor structure being moved forward to its new position without manual aid, to see the hydraulic roof supports lower from the section of the roof they were supporting and move forward by their own power to their new position and extend to meet their new position against the roof, and wondered at the revolution that had taken place.

It is now a common development either on the surface or below ground in the mining industry to find a multiplicity of operations, extending over a large area of the mine, brought visually into a control room through the medium of closed circuit television, with the man in the control room able to control and regulate several operations from a panel of lights, push buttons and graphs.

If this kind of progress has been possible in mining techniques, the progress in other industries and commercial undertakings where static physical conditions form the environment, where because of this it is infinitely easier to automate and computorize, the changes must be even more dramatic and revolutionary in comparison with the position some generations ago.

In the chapter of his book dealing with automation and labour, Dr Foster begins with the following observation:

'Labour is deeply concerned with the implications of automation. There is no use glossing over the fact that automation of certain industries may involve a net reduction in the number of workers employed. In other industries automation may so stimulate productivity and market outlets that the net effect may be to leave the total number of workers unaffected, but change the emphasis of the sort of work they do.' (Chapter 22, *Modern Automation.*)

This means that overall economic planning and control is required if adverse effects from this development upon total job opportunity is to be avoided.

It was Mr Harold Wilson who, when addressing the miners'

Big Meeting in Durham in July, 1963, estimated that some 10 million jobs in Britain would face dislocation as a result of this second revolution. He was not suggesting that current holders of these jobs would become unemployed, but was stressing the urgent necessity for central government to introduce economic and social planning to meet the consequences that would be thrown up.

Since 1964 Mr Wilson has been the head of the Government with the obligation to put into effect the planning measures he advocated in 1963.

While it can be argued that more needs to be done, it must be acknowledged that legislation has been introduced to take care of some of the more serious consequences of technological change. The Redundancy Act, the new benefits written into the Social Security legislation and the inducement offered to firms to take industry into areas to meet the need for new jobs, are some of the measures introduced to alleviate at least some of the worst effects. The criticism of the measures so far taken by the Government is that they do not go far enough; they alleviate but do not avoid evil consequences. This is the sort of criticism that can always be levelled against legislation in a free enterprise society that does not remove, but rather perpetuates the basic condition of original sin.

The ultimate solution to these and many other problems is in a fundamental transformation of society generally and there is certainly nothing wrong in agitating for this. Economic and social planning must be limited in scope and effect in a private enterprise economy. Effective planning requires a broader base, made possible by the extension of public ownership, at least in the vital sectors of the economy. The development towards this, if the pattern of the past is to be taken as a guide for the future, will be piecemeal and relatively slow. The march of events in the field of technology is proceeding more rapidly and although long-term unemployment is not substantially increasing there are serious problems affecting employment in many parts of the country. This is particularly true of areas where the older industries are contracting. In densely populated areas, built up for over a century to feed the mining industry with labour, the combination of contraction and technological change in the industry has created great hardship and social dislocation. It is not possible to accurately gauge the proportion of the dislocation due to technological change. But concentration and rationalization in the mining industry has been extensive and intensive and has undoubtedly contributed to the present problems in mining areas. From the standpoint of the remedies necessary, the precise cause does not much matter, but the effect of the technological revolution, as applied in mining, has had the effect – particularly with the older and the disabled men – of producing the situation forecast by John Snyder where, in places like Rhondda, West Durham and Cumberland, to cite only a few, conditions are not very different from his description of Detroit.

EMPLOYERS' ASSOCIATIONS

If there is any part of the industrial and commercial set-up in Britain that is more chaotic and disorganized than the employers' associations, both in industry and nationally, it must be in a complete state of anarchy. Employers' associations in this country are indeed in a sorry mess and in no shape to criticize union structure. The implication of this comment suggests that employers' associations need to be reorganized to some state of coherence and unity, at least in their outlook and policy in industrial relations. The implication is not accidental, because it appears sensible that, for as long as the present economic set-up continues, the employers' representatives in these associations should adhere to some common discipline and code of conduct in industry. Against this will be argued that if employers' representatives become better organized they will represent a more formidable body in collective bargaining, making it more difficult for unions to succeed in obtaining improvements for their members.

This kind of reasoning has its source in the belief that improvements are easier to obtain if employers in an industry act individually, or when plant management is given power to settle wages and conditions of employment. Generally, where this reasoning carries force and is based on experience, the trade union position is no better than that of the employers, and in fact can sometimes be worse. It can be worse because no matter how fragmented is the employers' association there is only one employer in the industry or plant, but there can be a dozen or more sections of the union. Although this obviously represents a weakness on the union side, the merit of unified organization on both sides has to be judged by its ability to achieve rationality in the wages system in the industry and in wage relationships between one job and another. It is axiomatic that division and disunity in employer and union organization invariably means a chaotic and incoherent wages structure.

The theory behind this belief is that better results can be obtained by unions from piecemeal bargaining, using the gain in one place to obtain a better one in another and so keep this step-ladder process going as a system of wage negotiations. This method, it is true, can give, where favourable conditions obtain, certain temporary advantages, but as a general procedure it usually creates more problems than it solves. The gains can usually be more easily obtained for key occupations or for the type of skills in short supply and less easily for the jobs that are non-specialist and where the law of supply and demand on the labour market is not so favourably disposed. The piecemeal, sectional process of wage movement, usually in the form of allowances or bonuses, responding to the erratic pressures of different occupational groups, results in disparities in earnings between one group and another, so that the relief of pressure in one occupation by this means creates it in another. And the process can go on *ad infinitum*. Contributing to the method, of course, is the existence of multiple unionism in these industries. But even where there is no such multiplicity of unions, the end product of the method is a pretty constant state of disaffection by one section of workers or other, which more often than not creates disunity in the general body of the labour force.

This, at least, has been my experience in the mining industry when these were the operating tactics; and it was the same under both nationalization and private enterprise. It led to a position where out of it we had the phenomenon of 'tit for tat' strikes. The walk-out by a section today meant others being sent home; and when those who first walked out went back to work, those who had been sent home walked out to spite the others – hardly a situation likely to promote unity in the ranks.

I am opposed to this method for reasons which for me are fundamental. The strength of trade unionism is in its ability to unite in a common policy or action the interests of all its members, or all the workers in an undertaking. It has to be seen by the members as a whole as not favouring privileged or economically strong sections and neglecting the less privileged and less strong. This is not just a necessary basis for policy in industry, on wages and other reforms; it is probably even more necessary and vital when related to the longer term objectives and aspirations of the trade union movement as a whole. My experience is that this unity can be dissipated and disintegrated by the internal strife and acrimony created by such tactics which in the end are usually for short term advantages only.

But to get back to the question of employers' associations, it is perhaps useful to quote from the *Donovan Report* (paragraphs 79 – 82):

'In some instances methods of Government and administration of the associations have failed to keep up with the times. The engineering Employers' Federation still has no formal place in its constitution for the large companies owning many factories which have come to play so important a part in the industry. The Federation's 4,600 members are grouped into thirty-five local associations. Companies cannot belong to the Federation directly but individual factories belong to local associations. Some large companies have both "federated" and "non-federated" factories. Consequently the General Council of the Federation, elected by the local associations, does not give adequate representation to the major firms. The real business of the Federation, however, is conducted by a Management Board made up of office holders, representatives of the local associations and co-opted members. The latter are "persons who from their wide industrial experience, can make a valuable contribution to the deliberations and discussions of the Management Board". The majority are the senior directors of major firms but most of them lack specialist knowledge in personnel matters. To meet this deficiency the officers of the Federation have to devise special means of consultation outside the formal constitution.'

This is the sorry state of affairs in the engineering industry and is true of some others. The reason for the existence of these Associations is equally doubtful when examined against the reasons they give and the practices they adopt. Paragraph 82 throws some light on this:

'Above all, however, membership of employers' associations is a consequence of unquestioning commitment to maintain the formal system of industrial relations. The system provides for industry-wide bargaining, employers' associations are essential for this; therefore companies belong to employers' associations. In practice managers have more and more difficulty in reconciling reality with the formal system, but most of them have not yet questioned the formal system itself. On the contrary, it still seems to offer an alternative to anarchy. It has stood the test of time.'

Well, it may have stood the test of time, but so have many other ancient monuments whose use value now is more aesthetic or historic than practical. This would seem to apply to the effectiveness and strange structure of this Employers' Federation. One can understand the difficulty of managers in reconciling reality with the formal wage bargaining system, which establishes wage rates that are only part of the final rates which operate and which are in the main determined at plant level. If this is the alternative to 'anarchy', it is because the employers' associations are so loose as to be impotent in securing the internal discipline necessary to a genuine and effective wage-bargaining procedure. In fact, this is admitted by the Commission in the following paragraph:

'The readiness of employers to federate does not, however, arise from a strong desire for strong organization. On the contrary, the changing status of employers' associations is the consequences of companies allowing matters to be settled in their own factories and workshops instead of holding closely to common regulation.'

Although the Donovan Commission could make this comment, which acknowledges that the prevailing wages system in the industry is the result of a failure of companies to 'hold closely to common regulation' and criticism of them is implied for not doing so, yet the recommendation they make is, in effect, directed to perpetuating the system of plant or company bargaining. However, more about that later. The post-Donovan decision by employers to review their organizational structure, will, it is hoped, radically change this situation.

I have previously tried to prove that it is possible to negotiate and settle a wages structure for the whole of a private enterprise industry which includes a number of separate companies. Two factors are necessary to facilitate this. One is that the trade union is able to negotiate for all the manual grades in the industry and the second, that the employers are organized in an association and have delegated authority to their representatives to act for them in discussion. This is not to claim that it is utterly impossible where more than one union is involved and the occupations to be included must go outside manual grades, or that a single authoritative association must represent the employer. But single organizations on each side obviously makes the task a lot less complicated.

Over thirty years ago there were negotiations with the South Wales Coalowners' Association to introduce a wages structure for all manual grades except those traditionally covered by piecework contracts. These were in the main men employed at the coal face, for whom the negotiations were to fix a minimum rate. The result was an agreement which became operative from April 1937, covering all timeworkers in the coalfield. The agreement provided uniform day rates to apply to all the pits. Jobs were classified into a number of grades on the basis of a rough process of job evaluation, and the rate for the job applied no matter in what pit it might be performed. During the period of the discussions, I was a member of the union's Executive Committee and was able to be present during the arguments. The union's case was being presented by the late Arthur Horner, a brilliant negotiator, against an employers' association that was as tough as they can come. But agreement was reached and applied until it was overtaken by a similar kind of national agreement in 1955 and such was the discipline on the owners' association that I can recall no occasion when an individual company refused to carry out the agreement or the interpretation and amendments to it made jointly by the parties through the conciliation procedure. If it was possible in 1937 with a collection of the toughest employers in British industry, as the inter-war years surely prove, it ought to be possible now, thirty years later, in circumstances vastly more favourable than existed then. The anarchy of employers' associations is clearly a major impediment to this, but there can be no justifiable reason why order in this limited field could not be achieved quickly, making possible industry-wide collective bargaining procedures capable of introducing and operating a rational wages structure, that is, if the parties themselves accept that such rationalization is necessary.

It is, however, not just a matter between them, but one that I am sure will more and more become the concern of the trade union movement as a whole and of central government.

THE STATE AND INDUSTRY

The state has always been involved in one way or another in the affairs of industry, ever since the first industrial revolution. But it has in the main kept 'aloof' from the processes of collective bargaining. This is how it is put in the *Donovan*

Report. Throughout the last century and through most of this one, State involvement has been mainly concerned with conditions of employment, particularly concerning safety, health and welfare. Many and varied is the legislation enacted throughout the years by governments that have have been persuaded or coerced into introducing measures designed to improve industrial conditions. Such legislation has obviously covered a wide range of subjects, among them being the more important reforms controlling the age of entry to industry; banning employment for women in industries like mining, etc.; control of deductions from wages; legal rights to compensation or damages for men who suffer injury; the right of employees to combine in trade associations to protect and advance their conditions of life; maximum hours of work for juveniles in certain occupations; and a lot more progressive measures too numerous to list here. The general effect has been a significant limiting of the harsh conditions of employment and an extension of curbs and restrictions against the more vicious forms of exploitation which characterized industrial conditions over a long period of history. This list of reforms is, of course, by no means complete and the more powerful trade union movement of today is constantly involved in discussion with the Government on a further number of important developments in this direction.

The trade union movement has naturally had its criticism of the Government in respect of the wide-ranging legislation already on the statute book. It is its proper function to continue to press for improvement in existing legislation and for the enactment of further legislation where it considers this necessary. This, however, is accepted by both Government and unions as a normal relationship, where the discussions are highly technical and generally conducted strictly according to protocol and in an atmosphere of moderation.

In the area of wage bargaining, the State has avoided direct involvement over the years, but this attitude of 'aloofness' has changed significantly in the post-war period.

Previously, their direct interest was with their own employees and in the work of the Wages Councils. Less directly they have been responsible for appointing disputes tribunals and aribtration machinery as a means for settling disputes where a failure to agree has resulted from the normal collective bargaining procedures. They have, too, intervened from time to time when strikes have been in progress, introducing and applying measures to maintain services, sometimes by the use of troops.

If the advantages from their involvement have to be measured by the wage rates that have been decided by the Wages Councils, then the outlook is anything but encouraging. The rates of pay for the workers covered by these Wages Councils are among the lowest in the country and are much lower than the rates negotiated through other industrial procedures. The *Donovan Report* makes the point that out of twenty-two wages council industries for which they had figures, only four had higher earnings than those in general industry and they were cutlery, paper bags, paper box and road haulage! The Wilson Government's announcement that it intends through future legislation to introduce measures to encourage voluntary collective bargaining in these industries is a welcome prospect and it is to be hoped that it will lead to the strengthening of trade unionism too.

As with all other things the role of central government in industrial affairs is changing. Their involvement has significantly increased in recent years and there are all the signs that it will continue to grow. This increased interest in detailed industrial affairs, including wage determination, is not peculiar to any one kind of government, as has been demonstrated by both Tory and Labour governments in the post-war period. In the modern economic and financial structure of nations, with the close interdependence of economics and especially finance, it is difficult to look ahead, to visualize a situation so changed, where this government interest will cease or substantially modify. It is not likely to change because of a change of government, the underlying conditions leading to this situation are not, it seems to me, government made, although government policies obviously can act for better or worse in relation to them.

Three important questions can be posed, the answers to which might help clarify the position.

What are the changed conditions that have brought about this State involvement in wage determination?

Is it a position peculiar to Britain or is it the same in other countries?

Is it a condition that is here to stay?

Not possessing any specialist knowledge of economics and certainly not of high finance, I can only offer answers within the limits of my own understanding, drawing a bit from experience and applying what common sense I possess. In earlier pages reference has been made to the changes taking place in the form and size of organization in industry and commerce. This, together with the tough conditions of modern trade, the intensity of competition, the drive to cut production costs, represents a continuous driving force to more rapid concentration and centralization, demanding ever larger capital investment as a condition for economic survival. Add to this the second industrial revolution and the massive cost involved in reorganization of production systems, and some idea of the dimensions of the economic and financial problems arrive.

To lay out a modern, semi-automated coal face in an average length of face, equip with modern machinery and supports, the required initial capital outlay is in the region of a quarter of a million pounds. Equip a single colliery with four or five such coal faces and the total sum is over a million pounds. The provision of power, materials, maintenance and labour and the operation represents big money. The same position applies to the construction of an electricity generating plant and more so to one based on nuclear power; to building a modern oil tanker, or a substantial factory building; the construction of a major road or a block of flats; the construction and installation of a computer complex or the equipping of a modern farm, in fact any kind of modern development. The capital required is so huge as to be beyond the capacity of even the larger companies, so they need to borrow and obtain credit and the necessary backing in order to get the essential capital.

The whole business is very complicated and to a large extent shrouded with the mysteries of high finance. A take-over bid involving nearly a thousand million pounds is not just money from the company safe; it obviously has to be raised by the various means available and to raise it security has to be offered. So the banks, the big insurance companies, money-lenders of all kinds, inside and outside the country, are drawn in, attracted by the interest their loans can earn. But at the back of it all, because it is part of the financial structure of the country, the Government is the final security. Its financial stability, the state of trade, the health of the economy and so on – these are the factors needed to give the investors confidence in the big projects of construction and reorganization. The Government is consequently involved directly in maintaining the stability of the nation's financial activities, as well as its international standing. No part of the financial scene can be ignored and that, the Government argues, includes the movement of money in the dealings between employers and employees.

The second big factor which draws the Government into some surveillance of what is happening in industrial relations, is the social obligations for which it alone becomes responsible as the technological revolution gathers pace. The effects are not only within industry, they are also social and that in a big way. As we have said earlier this is the age of automation, the substitution not only of man's physical labour but an encroachment on his mental contribution too. The electronic brain is a fact of life and is the means that will create perhaps the biggest impact upon job availability. This is the device that will not only revolutionize production systems, but will probably have its biggest impact in the area of operations occupied by the supervisor, the controller, the administrator and those jobs usually designated 'white collar'. The following is a description of this new process extracted from an article published in an American magazine back in 1965. It adds to the descriptions already given:

'In contemporary plants manufacturing automobile engine blocks, for example, the crude blocks are mechanically *transferred* between machines which automatically perform the various broaching, boring, milling and other *processing* operations. Automatic inspection instruments are used, and, in some instances, feedback devices automatically *control* quality of output.' (Author's italics.)

These devices operating with machines not only determine the quantity of production, but also the quality of the product. The effect is not only to reduce the labour required, and this applies to offices as well as the whole process of production, but affects the nature of the jobs required by the new system.

Mechanization simplified jobs and created a much greater subdivision of the jobs in a given operation. With mechanization as we have previously experienced it, the trend was to specialization on a single phase of an operation. One of its by-products was a big increase in occupational tedium, created by the repetitive nature of the job. Plenty has been written around the measures, gadgetry and gimmicks employed to relieve the monotony of these repetitive tasks. I remember reading one account years ago of a firm employing girls on this single purpose operation, using a cat along the long bench or table where they worked, so that smoothing the cat had the effect of relieving nerve tension and at the same time relaxing the finger muscles so that they maintained their efficiency in the selective work they were engaged upon!

The process of job simplification was, one supposes, a necessary pre-condition to the application of automation, where each single and special operation is performed automatically, producing in total a whole complex of automated operations. The effect means that the subdivision of labour created by what now may be called simple mechanization, is being overtaken and replaced by the concentration of a whole operational process into an automated complex requiring only the attendance of monitors for the control panel and technicians to maintain and repair the equipment. This means that former skills and job expertise can also be made superflous and redundant. Thus new job names and new job skills and expertise peculiar to the new techniques arise and must in the end reduce the number of job names as well as the number of jobs, certainly as compared with the previous subdivision of work. It has at least one beneficial possibility in that it will make it easier for those who may want to initiate a rational wages structure to sort out job classification and to slot them into the appropriate grades for the purposes of wage payments.

It will also lead to the need for a new type of skilled craftsman; one specially trained not only in electronics or mechanics, but one trained to a high specialist knowledge in electronic, hydraulic and mechanical engineering. Within craft sections there are already signs of an attitude of opposition to this change in the time-honoured conception of their craft. For some time there has been great difficulty in getting accepted the combined craftsman, the electro-mechanic, a skilled man trained and competent to diagnose, repair and maintain machinery in which both factors are integrated. This opposition to any widening of the traditional definitions of craft are deep-rooted and the transition to a new attitude is likely to be long and difficult. Training in these new skills, while being a responsibility for industry, must also be incorporated in the training schemes of the Government and as part of the whole concept of central planning.

The effect of the new techniques is massive and shatters many hidebound attitudes and stances that grew up with the previous production systems as well as disturbing the traditional way of life for families and communities. Where automation operates, the 'who does what' demarcation issue has no basis. As in the example already cited in the construction of motor-car engines, the whole process is undertaken mechanically and automatically, the completion of one specific operation engaging the mechanism for the next. The cumulative effect of these changes must lead to changes in wage relations and to demands for wage increases where the content of jobs change because of the new process and where fewer men are required within the operation. There are plenty of examples where the introduction of push-button controls or automated or semi-automated devices have halved the labour previously employed in an area of operations. The reduced labour force will want a share of the saving in labour costs involved. This is a legitimate and understandable demand, but whether the benefit should be in some wage increase to the section of men directly affected, or used as part of a general case for increases for the whole of the labour force employed in the undertaking, can be a problem difficult to resolve. The effect of applying such benefits to a section creates disparity in the wage relativities as between job and job and usually leads to upsets between men on the job. Since the process of change is general it would seem more just and equitable that the benefits flowing from improved productivity should have general application.

The experience already exists in a large number of industries where the capital outlay to construct and equip the plant requires its constant operation of twenty-four hours a day and seven days a week. The recovery of the capital cost, plus expected profit, is assessed against depreciation of the plant itself. Machines wear out and buildings have to be maintained like everything else; they have in simple terms a 'life expectancy'. To realize on the original capital outlay over the expected life of the plant can mean continuous production. This is not a new feature, but is one that can only extend as the new technology advances to all aspects of economic activity. More industries and more operations ancilliary to industry will go over to continuous production. Not only is it considered necessary because of the huge capital cost of the new processes, but also because maximizing the running time of machinery improves productivity, and by this, lowers costs of production. In many industries now, the drive to increase productivity is concentrated as much on increasing the time of machine performance as on the application of new techniques, the new techniques, in fact, accentuating the necessity of extending machine running time.

Our social habits and the availability of social amenities have been traditionally geared to evenings and weekends.

Theatres, cinemas, sport, entertainment of all kinds, are still conditioned to a way of life that is disappearing. Even television offers nothing for the mornings and early afternoon. It is true that some social clubs in certain industrial centres are beginning to cater for the shift workers. But generally the position is geared to a leisure-time position that might have applied a hundred years ago.

A large part of modern industry is already operating continuously and this will grow into predominance. Shift working, working weeks that include weekends based on five shift rotas, sometimes changing shifts at shorter intervals than five, this is the order of modern industrial operations. The effect on family and community life is pretty obvious. Not only are the breadwinners of families working these staggered shifts for a large part of their working lives denied opportunities of taking advantage of the social amenities available, but the whole family life is affected. One of the benefits that must result from automation and technological change is surely an increase in leisure time, by reduced hours of work per shift and a shorter working week.

The social consequences are not confined to redundancy and the need for new industry; the big impact is on the blackout imposed on the ability of people to take part in the normal and traditional forms of social intercourse. This aspect of the problem is not being attended to at anything like the speed or seriousness warranted, and represents yet another responsibility which can only be tackled on the initiative and planning of central government. More than ever before the need for economic and social planning is of paramount importance, not only to provide for job security and job opportunity, training and retraining, but to adapt cultural and entertainment facilities to meet the new order of leisure time to which millions of workers are subject now and will be in the future.

It has already been acknowledged that some progress has been made through legislation, but a lot more needs to be done. However, my point is that the main responsibility for accommodating them rests on the Government. And to discharge those responsibilities must inevitably involve substantial financial cost, which the Government must meet and this obviously forms another important consideration in the totality of Government money problems and their control.

The new industrial revolution could not have got under way at all without the direct involvement of the Government in its promotion and financing and in meeting the industrial and social consequences it creates.

In any study of the Government's intervention in industrial affairs there is, of course, another factor relevant to the argument. The Government is a big employer of labour and is directly involved in industrial affairs as an employer. In this role it is directly involved in wage determination. Apart from this it is, as has been already shown, a big investor directly and indirectly. Its investments are widespread, covering housing, education and science, social services, transport and communications, economic development covering trade and industry, agriculture, forestry and fisheries, as well as government buildings, etc. The increase in investment in this range of Government activity has been colossal in recent years and creates a compelling interest in the ways these monies are actually used and this inevitably can include industrial relations.

It is against this background of compelling economic and social obligation that the direct involvement of central government in the affairs of industry, with special concern and interest in the factors that affect the country's financial stability, i.e. prices and incomes, must be judged.

INCOMES POLICY

So far this study has been mainly concerned with the functions of the State in industry in a general way, concerned with policy direction and supervision of the economic and social effects of the changes taking place. This is not the aspect of their present function that has produced within the labour and trade union movement a spate of criticism and opposition. It is the Government's intervention through a legislated incomes policy and proposals for trade union reform and the curbing of unofficial strikes, that has produced the criticism and opposition.

It is probably advantageous in considering this controversy to distinguish first between incomes policy as a general political principle and the form it may take in particular conditions; the form it takes, for example, in the Government's present scheme. My own view, based on the reasoning already made in earlier pages, is that an incomes policy should operate. It is extremely difficult to conceive how central government planning – and the need is for planning far more extensive and effective than is now the case – can operate over all other economic activities yet be excluded from incomes. This is not a defence for the form of incomes policy now being applied, but is intended as a defence for the principle. An incomes policy must be something more than just wage restraint, it must apply to all incomes and must be part of a plan aimed at greater equity in wealth distribution.

My own view is that we have reached a position of such close interdependence, even integration, of world finance, that governments in all countries are compelled to be more directly concerned in the control of money in all aspects of its use, whether in wages, salaries, profits, dividends, rents, prices, investments at home or abroad. The concern of every country is to achieve and maintain stability of its financial structure. Since the end of the last war this country and others have faced recurring crises due to adverse balances of payments and the financial problems that this creates. My assessment of the causes are along the following lines, and I start with an acknowledgement that it is a capitalist society we live in and from which these problems arise.

First, we are a small island with a relatively big population. Our natural resources of wealth – and we do not use very wisely those that we do not possess, as witness, coal – are limited and to exist we must trade the products of our industry for the things we need to live.

Second, we have the resource of being highly industrialized, although, compared to other leading industrial countries, lagging behind somewhat technologically, and we possess a highly skilled and experienced labour force. As distinct from the time when we dominated the countries from which we obtained raw materials for our industry, we now have to buy them at world prices. Third, we have in addition, commitments in various parts of the world and to other nations, whether justified or not, involving financial obligations that we are expected to meet.

Because of this combination of factors, the wealth we produce has to meet the cost of the things we import as well as our overseas commitments and our internal responsibilities in respect of industry and people. An important element in the fulfilment of these liabilities is the vital necessity for our exports to other countries to earn enough to meet the cost at least of the things we import. This adds up to the matters involved in a national budget and reveal whether we are solvent or not. If we have an adverse balance of payments and we produce insufficient wealth to meet our liabilities, we borrow from the World Bank or other international banking houses to meet our deficiencies. If deficits in national accounting continue, we continue to borrow and have the added liability of repaying the sums borrowed in a stipulated time with the interest charged. There is quite obviously no future in continuing on such a course and sooner or later borrowing becomes more difficult and the conditions attached more tough. This was, in fact, the substance of the speech made by Mr Harold Wilson to the Trades Union Congress in September 1964, on the eve of his becoming Prime Minister:

'. . . we can borrow from the International Monetary Fund, from the United States who have placed large sums at our disposal and by swap arrangements with Central Banks.

'But we can't go on in this humiliating posture – having to borrow everytime we dare embark on a timorous little policy of expansion and fuller utilization of under-used industrial capacity. Yes, we can borrow, that's where thirteen years of Conservative rule has brought us. You can get into pawn, but don't then talk about an independent defence policy.'

Five years have passed since that speech was delivered, five difficult years during which we have regrettably continued borrowing, and have devalued our own currency as an added measure to find stability, but the truth in 1964 is the same in 1969. Such measures are not solutions. This is just one layman's brief explanation, not justification, for what obviously is a very complicated combination of factors. It could not hope to be comprehensive or authoritative and is not offered as such.

These crises arise from the structure of international capitalism and its economic and financial relations and although this sounds profound, it could not very well be anything different, society being what it is. The remedy to all the problems created by capitalism is socialism, this is indeed the faith of every socialist. It is my faith. Some forty years ago I thought socialism was just around the corner for Britain. Unfortunately it is a very elusive goal and difficult to reach, because it has to be reached by the action of people. Posing socialism as the alternative is right as the ultimate, long-term remedy, but it is no more than an abstract slogan as an immediate alternative. However, the policies we pursue as the immediate alternative must dovetail in with this ultimate strategic aim of the labour and trade union movement.

The problem is one that has to be faced by government, no matter what its particular creed or aims. It seeks to control incomes as part of a policy to limit spending as a means of reducing the import bill and restoring a satisfactory balance of payments. There are many other aspects of national spending that can be cut and part of the criticism made is that the emphasis of the policy is directed against consumer spending by limiting wages. The device itself is anything but new, but it is, as history shows, a device that can be used more effectively by one kind of government than another. Conservative governments have had great difficulty in operating restraints on wages and other incomes. Labour governments have been a little more successful, but in the past, their success has been comparatively short lived.

In 1947, the labour and trade union movement faced a

situation not unlike the one it faces now. The Labour Government elected in 1945 faced a serious financial crisis aggravated by an acute fuel famine. The appeal was made to the unions to operate strict restraint in respect of wage claims. I can remember the heated debates inside the conferences of the Miners' Union and the divisions that were created. It is worth recalling in more detail the situation that led to this appeal and the response of the trade unions to it. What is immediately apparent, however, is that we again travel the road over which we travelled more than twenty years ago and the measures we accepted then as remedies have not prevented a recurrence of the complaint. The Labour Government had been elected to power in what were obviously difficult times. The havoc and devastation of war had to be repaired, industries and towns reconstructed. Alongside this a programme for major social reform was started, the foundations of the Welfare State laid and nationalization of important industries was effected.

It was a period of extensive financial aid being made available from America through what was popularly called 'Marshall Aid'. Britain had received a big American loan in 1946, but with rising prices, American imports requiring dollars, plus the effect of the fuel crisis on the country's ability to export, created the conditions for a serious financial crisis. In fact, the position was so serious that in the early part of 1947, unemployment almost reached the two million mark. The end result was a heavy balance of payments deficit and a state of inflation, a condition where 'too much money was chasing too few goods', as the popular definition of inflation puts it.

To meet the fuel crisis aspect of the problem, the miners, although just having been conceded a five day working week, agreed to working Saturdays to boost coal production, but any benefit from this was only limited and of little immediate consequence.

Early in 1948 the Government issued a White Paper on *Personal Incomes, Costs and Prices* which put the Government case for incomes restraint. The case then was largely identical with the case put before and since. Put shortly, it stated that, to avoid a dangerous inflationary situation, personal incomes should not be increased unless there was an increase in the volume of production. The Government's appeal was considered by the General Council of the TUC who recommended, with some reservations, that the principle of wage restraint should be accepted. This recommendation was put to a special Conference of Executives of Unions in March 1948 and was carried, although there was criticism and opposition from some of the unions.

The situation, however, deteriorated, with the result that, after further discussions with the Government, the TUC dropped the reservations they had previously held and accepted the Government's position completely. The policy of the TUC continued in operation until the Congress in September 1950, when by a majority vote, despite the advice and intervention on behalf of the General Council, it was defeated. Some aspects of the TUC position during this period gave rise to a great deal of cynical criticism when many of the leaders who supported restraint were in the front of the queue for wage increases in their own industries. The foregoing is a very brief summary of the history of the incomes policy from the standpoint of procedures within the TUC. The detail shows the similarity of approach both by the Government and the TUC, then and now.

On February 4, 1948, the Prime Minister, Mr C. Attlee, made a statement to the House of Commons outlining and explaining Government policy. This was subsequently issued as the White Paper. A summarized version of the statement asserted that increases in personal incomes, without corresponding increases in production, would lead to inflation and our products would be too expensive to sell abroad. It said that some success had already been achieved by high taxation of incomes and profits and that food subsidies, by keeping down living costs, had checked demands for wage increases. There may have been some instances where wage increases could be justified, such as manning up a particular industry or where levels were very low. Other incomes such as rent, interest and profit should not increase but there was no freeze of wages or interference in the processes of normal wage bargaining!

This is very similar in content to the aims set out in the White Papers in recent years, with the same kind of exemptions, the main difference being the legislation that now applies, giving incomes restraint, a legal framework.

The TUC comment on the White Paper is set out in their Report for 1948. It stated that the policy of the White Paper was acceptable to the extent that it would meet their reservations which they considered essential and would:

(1) Recognize the necessity of retaining unimpaired the system of collective bargaining and free negotiation;
(2) Admit the justification for claims for increased wages where those claims are based on the fact of increased output;
(3) Admit the necessity of adjusting wages of workers whose incomes are below a reasonable standard of subsistence;
(4) Affirm that it is in the national interest to establish standards of wages and conditions in undermanned industries in order to attract sufficient manpower; and
(5) Recognize the need to safeguard those wage differentials which are an essential element in the wages structure of many important industries and are required to sustain those standards of craftsmanship, training and experience and contribute directly to industrial efficiency and higher productivity.

This was quite a charter of reservations, but as will be seen some were already conceded. There is, however, one important aspect of these reservations that is not conceded in the present policy either by the TUC's voluntary system or that of the Government. This is especially true in respect of industries with wage structures based on job evaluation. The imposition of a ceiling in general increases, with relaxation of the ceiling where the increase applies to lower paid, can have the effect of telescoping those structures and destroying the established differentials. Neither the incomes policy of the trade union movement nor the Government has within it any perspective of creating reasonable wage structures embodying some worthwhile principles such as a uniform rate for the job, some equity in differentials within particular industries and between industries. The main content of both is wage restraint.

The other condition, in 1948 as compared with the present is, of course, that the Government made it clear and took no direct steps to interfere with actual wage bargaining. The TUC adopted the same attitude leaving it to the individual unions to interpret 'restraint' and this, as could be expected, led to a variety of interpretations in the practice.

In September 1949, the pound was devalued in relation to the dollar. The TUC in its consideration of this development took the view that if the country was to get the benefits from this improvement in our competitive position as a country, we had to prevent export prices rising. They urged unions, as one means of ensuring this, 'to exercise even greater restraint on wage increases in the future than they had in past'. (TUC *Report* 1950)

The General Council called upon all affiliated unions to 'pay regard to the realities of the economic situation in framing their policy and to act loyally in conformity with the policy now recommended by the General Council'. The policy recommended was for 'rigorous restraint'. The view of the Council on the situation was set out in the first paragraph of their proposals to the unions:

'(a) The General Council consider they should once again stress the gravity of the present economic situation and emphasize the fact that devaluation has been adopted as an alternative to deflation. The dangerous inflationary tendencies which devaluation inevitably intensified must be counteracted by vigorous restraints upon all wages, salaries and dividends.'

The General Secretary, Mr Vincent Tewson, led in for the General Council when their report was being considered in the 1950 Congress and had the job not only of justifying the General Council policy, but also of rebutting a resolution contrary to the Council recommendation and the policy which had operated since 1948.

The resolution was moved by the then Secretary of the ETU, Mr Walter Stevens, and its central theme was in its middle clause:

'Congress is of the opinion that until such time as there is a reasonable limitation of profits, a positive planning of our British economy and prices are subject to such control as will maintain the purchasing power of wages at a level affording to every worker a reasonable standard of living, there can be no basis for restraint on wage applications.'

Summed up in that clause is the main content of the argument against wage restraint; it is the burden of the case presented at the present time by the unions and Congress against current wages policy. Indeed, it is on the records that at a Special Miners' Conference, in December 1949, I spoke very strongly against the wage restraint policy, the last sentence of which was that 'we should see in the crisis of capitalism an opportunity for socialism to advance'. The theory underlying this statement is that out of the crisis and collapse of capitalism, socialism will emerge, the logic being that the crisis of capitalism should be deepened and pushed toward collapse. This theory is sometimes expressed in another way, by contending that it is not the job of the labour and trade union movement to make capitalism work. The opposite way of putting the same argument is to say that it is the job of the labour and trade union movement to stop or hinder capitalism from working. How much of this thinking colours the outlook of trade unions at the present time is worth some examination and this is an aspect of policy that is best considered when this study deals with problems of trade union strategy in relation to modern conditions.

The discussions that took place during the period of the post-war Labour Government have been presented at some length and, whatever other conclusions may be drawn, there is one that stands out. It is that the nature of that crisis was the same as the present one and that the policies advanced to deal with it are the same. It also stands out that such policies do not stop such crises recurring.

During the years that followed, continuous efforts were made to get the unions to apply moderation in making wage claims. Between 1951 and 1964, the efforts were being made by a Conservative Government, but the trade union movement has always found it easier to oppose policies from that source than from Labour governments. The efforts were consequently largely confined to exhortations and formal representations, having little more response from employers than from unions. During Mr Thornycroft's period of office as Chancellor in 1957, the Government took direct measures to hold back wage increases to national service employees, civil servants and the employees of public corporations. In October of the same year the Chancellor had stated in Parliament that 'wage increases unrelated to, and going far beyond, the general growth of real wealth of this country, are by far the greatest danger we have to face. . . .' This, of course, is another way of saying that wage increases would have to be held back to arrest inflation, and this the Government did, wherever they had power to directly intervene, usually where they were the employer and where the trade unions were not strong.

At about this time, too, a Council on Prices, Productivity and Incomes was set up under the chairmanship of Lord Cohen, which from time to time issued reports dealing with the nation's economic situation, although it had little impact on the course of events, particularly in relation to wages and salaries, because the Trades Union Congress would not associate with it and regarded it with extreme disfavour.

Later, in 1962, during Mr. Selwyn Lloyd's Chancellorship, the efforts of the Conservative Government were stepped up. A White Paper was published on 'Incomes Policy'.

One of its proposals was that wage increases and increases in other incomes should not exceed a limit of 2½ per cent. It was anticipated that this limit would regulate such increases with the anticipated increase in productivity. They applied this policy wherever they could, and actually refused to implement awards providing wage increases to their own employees and opposed increases agreed upon by wages councils. Policy had definitely toughened and Arbitration Awards and Wage Council decisions were set aside by direct Government intervention.

Together with this sterner effort to curb wage increases, very strong pressure was also being exerted on employers to concede no reduction whatever in hours of work. There had been a breakthrough to the forty-hour week by a section of the building industry in Scotland and this had produced very strong pressures from other sections of industry, and there was some evidence of a disposition by some employers to agree to this reform. We had the backwash of this opposition of the Government to shorter hours in the mining industry. The union, ever since 1926, had agitated for the restoration of the seven-hour shift for underground miners. A breakthrough had been achieved through an arbitration award which resulted in a quarter-of-an-hour reduction. This was followed by an agreement with the National Coal Board that, when output per manshift reached 34 hundredweights, the further quarter-of-an-hour reduction would be introduced. This was a firm agreement. However, when the output target was reached in this atmosphere of Government resistance to shorter hours in industry generally, the National Coal Board shifted their position by proposing the substitution of seven rest days spread over the year for the reduction in shift hours. The areas of the union saw this as a more attractive reform and accepted it. The politics of the substitution was quite clearly dictated by the desire that the stand of the Government against reduced hours of work should not be weakened, especially by a nationalized industry to whom they could give directions. The seven-hour day has still not been restored in the form it was traditionally sought and most of the rest days are taken on an individual basis with a substantial proportion of the individual and pit rest days taken on days that would not be worked anyway.

In February 1962, the TUC accepted an invitation by Mr Selwyn Lloyd to join the National Economic Development

Council. The Council would be presided over by the Chancellor himself; there would be two other Ministers, who, together with representatives of unions and employers, would comprise the Council of about twenty people. Its purpose as explained by the Chancellor was to 'establish new and more effective machinery for the co-ordination of plans and forecasts for the main sectors of the economy'. The aim was to involve both sides of industry in plans to achieve more rapid and sustained economic growth. The General Council had many meetings with the Chancellor before they agreed to join. The decision was made difficult because the discussions coincided with an announcement by the Prime Minister reaffirming the Government's pay-pause policy. The continuation of the pay-pause policy was undoubtedly an embarrassment for the trade unions, but they finally gave way following further discussion with the Chancellor. His argument, as presented in the General Council Report to Congress in September 1962, sums it up:

'The Chancellor argued that it was necessary to distinguish between the short term and the long term problems of the economy. He recognized that in the short term problems the General Council and the Government were at loggerheads. The establishment of NEDC was an approach to solving the long term problems and, in this connection, he had been influenced by the TUC's views on planning. He genuinely wished to see the TUC participate in the Council's work. The General Council's representatives pointed out that the issues could not be confined to a disagreement over the Government's short term policy. The TUC had made valid criticisms of the Government's policies but these had been regularly ignored. The Chancellor said he was disturbed by the effects of the pay pause on industrial relations but he still believed the policy was necessary. The sooner the long-term planning got under way the sooner the policy of restricting demand could end and he hoped that the TUC would join the Council, which would enable it to have more influence on Government policy.'

This started a policy which has continued where the TUC has directly participated in organized discussion with the Government and employers on economic policy and planning. Previously, such discussions took place following a request by the TUC for a meeting, where it put in a more detached way its views on Government policy based upon its own studies. The issues of incomes policy is excluded from the discussions of the Economic Development Council, as a specific question, although it must obviously be involved in any serious consideration of economic growth.

In July 1962, the General Council was invited to meet the Prime Minister and other Ministers, to be told of an announcement that was to be made to Parliament later in the day. They announced their decision to set up a National Incomes Commission. The purpose of the Commission was to consider important matters which could be referred to it by the agreement of the negotiating parties. It would inquire into the circumstances around important wage claims and judge them from the standpoint of their effect on the national interest and whether they could be matched by increases in production. These were briefly their terms of reference. The Commission would express its view on the claims it investigated and, although it would not have power to alter settlements, its purpose was obviously intended to strengthen the restraints already being operated by the Government. The General Council were strongly opposed to the Government's proposal, and disagreed with the reasons put forward by the Government in support of their decision. These were the main points made in the General Council's condemnation, and are set out in an appendix to their report in 1962 on the National Incomes Commission. Summarized, their objections were: first, they did not accept that wage restraint was necessary as an anti-inflationary policy and they did not agree that Britain was drifting into an inflationary situation; second, that the Government was perpetuating old policies under a new guise and trying to shift the antagonism to such policies away from itself; third, the Commission could not define 'national interest' without arbitrary decision on a number of very contentious factors or acting as an agent of the Government; and finally, that the Commission was bound to have a bad effect on industrial relations, free negotiations and arbitration awards. For these reasons they refused to have anything to do with it. The National Incomes Commission stood condemned, but the National Economic Development Council was acceptable and of some possible use, although the TUC has demonstrated no great enthusiasm for it, then or now.

This is the story, then, of the evolution of incomes policy and wage restraint unfolding itself in the post-war period, from the primitive and shaky beginnings to the more robust, but still alien, body of the present – from a beginning based on exhortation and persuasion to voluntary restraints, to the formal and legally based machinery operating now; from a National Incomes Commission without teeth to the present Prices and Incomes Board with some teeth; and where the trade union movement itself accepts the principle of restraint and has created its own machinery as a means of applying it.

The conclusion to be drawn from this is that the incomes policy, which has aroused so much criticism in the trade union movement and against the Labour Government, is the continuation of a growth that has been with us for more than twenty years. It is a policy that the trade unions have adopted and applied to wages in varying degrees of intensity throughout the period, so that unions are pretty well committed to the principle although they continue to oppose statutory enforcement if voluntary restraint is not adhered to. The issue is not a new one and because it is not, a study of policy is important if there is to be some end to two decades of controversy around it.

THE STATE AND WAGES DETERMINATION IN OTHER COUNTRIES

The other question posed was whether other capitalist countries were experiencing a similar state of affairs. Generally speaking they are, although the degree of government intervention varies with the health of the respective economies and their tradition. In actual fact until recently the position in Britain compared to a number of other countries has been unique in the limited intervention by the State. Because it happens in other countries is of course no justification for it being applied here. The attitude of the trade union movement to this intervention also varies, and is influenced to a great extent by the conditions preceding it.

For instance, in the United States of America the State intervened with legislation which laid down collective bargaining procedures, and union recognition was part of the New Deal policy promoted by Franklin D. Roosevelt in the middle 1930s. Because of the repressive anti-union attitudes which preceded this legislation, American unions welcomed the intervention of the State and the Law into industrial relations. In West Germany the principle of legally binding contracts dates back into history and is accepted as the normal condition for agreements between employers and unions. In addition, there is of course, the background of Nazi occupation and the policies of regimented industrial relations with which they were identified.

In the following brief account of the position in the various countries it will be seen that although the form of Government intervention and influence differs, the aim being pursued is the same.

In the United States there is fairly comprehensive legislation controlling industrial relations. Contracts made between employers usually for three year period, are legally binding and invariably contain a 'no strike' clause. The Government has the power to impose an eighty day cooling off period in strike situations. The *Sunday Times* on January 26, 1969 contained the following information on this position:

'In twenty-two years there have been twenty-nine investigations ordered by the President which precedes the use of the cooling off period. On twenty-five occasions, the eighty day injunctions were introduced, and in fourteen of these instances the dispute was settled in the cooling off period.'

The Labour Laws provide for the enforcement of union recognition by employers and for one union being granted negotiating rights usually on a plant or enterprise basis. There is a National Labour Relations Board which ajudicates on certain types of labour problem. Most strikes are official and often are in breach of the legally binding agreement, with the unions being prosecuted and sued for heavy damages. This aspect of the Labour Laws is considered a paradise for lawyers.

Most of the unions in America are organized on a vertical industrial basis and the inter-union problems are fewer than here.

In France there has been for some time past a legal framework for wage determination and, although it still exists, its main function now is to fix minima, to which additions are made at enterprise level. The main pressure from the Government is directed to holding back wage increases in the public sector as a pattern basis to be followed by the private sector. In 1964 a special committee was set up and the policy they recommended to control pay increases, and to restore the financial position of the nationalized industries, was that the Government should fix the total sum by which the salary bill of the industry could be increased leaving the actual allocation of the sum to collective bargaining. In 1965 increases in nationalized industries were restricted to 4 per cent and were kept under 5 per cent in industry generally.

In West Germany anti-strike laws exist but are rarely invoked. Collective agreements are legally binding which means that strikes are illegal until the expiry date of the collective agreement has passed. Strikes can only be called after a secret ballot showing a 75 per cent vote in favour. There is a strong traditional discipline within the German Trade Union movement with power being very strong at the Trade Union Centre. Strikes are rare and occur only after negotiations have been exhausted.

The form of control in Belgium is less direct. There exists a network of economic advisory committees representing unions and management in each branch of industry whose chief concern is with the economic position of the industry and in this way management and unions are committed to participation in the formation of economic and social policy. At national level there is a series of joint bodies involving the unions which advise the Government on economic and social questions, and a National Council which establishes Conventions to settle for a period ahead general conditions of work, hours and holidays which apply throughout industry.

In the Netherlands and the Scandinavian countries, wage bargaining is centralized and is undertaken by the trade union central organization and the central organization of the employers. In the Netherlands the pay settlement has to be approved by the Government. The settlements determine both the maximum and the minimum industrial wages. There is a large measure of co-operation between unions and employers through a joint organization called the 'Foundation of Labour' which functions as an advisory body to the Government. If a disagreement arises between the Foundation and the Government on wages policy the issue is decided by State mediators.

In Denmark the position is relatively straightforward. Government policy in general terms is transmitted through a National Economic Council representative of the Government, employers and unions, whose terms of reference take in both prices and incomes. The system of wage bargaining is centrally undertaken and should there be failure to reach a settlement the issue goes to State arbitration. The arbitrators submit to the parties proposals aimed at securing a settlement. If their proposals are not accepted it is open to the Government to introduce special legislation to enforce acceptance and this has occurred on a few occasions.

In Norway and Sweden, the centrally negotiated wage agreements provide minima to which adjustments can be made at industry level. Although the Government is not directly involved in wage determination, negotiations between employers and unions are usually preceded by the parties being made aware of the views of the Government. An important factor in both these countries is the very close attachment that exists between the unions and the Social Democratic Governments.

The employers' organizations are subject to strict discipline from the centre. This is the position in Denmark, Finland, Norway and Sweden, where the rules of the employers' associations forbid members to conclude collective agreement without approval from the centre. On the workers side, although such powers are a little weaker, the individual unions accept a large measure of control from the centre. In Norway member unions cannot conclude agreements or take strike action without the approval of the Central Trade Union Organization.

In both Norway and Sweden the central trade union organization holds large central disputes funds, but in Sweden the central organization will not finance strike action from this fund in support of wages increases in excess of those laid down in the central agreements. The combination of all these factors exerts a continuous control on the processes of wage determination.

The position in Austria is controlled by a Joint Wages and Price Commission. This Commission is made up of four representatives from each of the three main centres, the Government, the Chamber of Commerce and Industry, and the Agriculture and Trade Union Federation. All proposals to change collective agreements are referred to the Commisison and their decisions are usually accepted. The claims that go before the Commission are defended by the Federation of Unions who act in these matters for the individual unions involved. Here, as in most of the countries referred to, union structure is vertical.

It is clear from this presentation of the situation in other countries that the methods of seeking to control wage movements vary. Most countries have a National Economic Advisory committee, representative of the unions, employers and Government, and although not all of them are directly concerned with incomes, they exercise an indirect influence upon this aspect of policy. In most countries too, the Government has some kind of reserve power to meet difficult situations. Incomes policy is now an established anti-inflationary policy device in use in all countries in one form or another when the occasion demands. The only merit of this brief recital is to put the problems and policies confronting the trade union movement in Britain into perspective as part of a world development. The intention is not to justify the policies being operated here or elsewhere but to establish their universality.[1]

IS THIS FORM OF INCOMES POLICY PERMANENT

The last question posed at the beginning of this section was whether intervention in wages determination by central government had come to stay. I can only offer a personal opinion and I think it has come to stay. It is an opinion based on an assessment of the factors dominating the economic and financial state of the country which have been mentioned in some detail in this review of relationship between the State and industry.

The important factors that support this point of view put briefly are as follows but not set out in order of importance. First, in this country and most other countries, in some form or other and for differing lengths of time governments have influenced wage increases by legal means. Through the whole of the post-war period central government in Britain has continuously pressed for control of incomes and especially of wages, the trade union and labour movement accepting control in principle but objecting to legal sanctions and compulsion.

Second, the development of giant economic organization in industry and the promotion of technological change calls for huge capital investment and expenditure, which the Government either has to provide or underwrite.

[1] This information on the position in other countries has been culled from two main sources, *Prices, Wages & Incomes Policies* by H. A. Turner and H. Zoeteweij for the ILO, 1966; and *Incomes Policy in Post-War Europe*, a series of reports to the EEC.

Third, the size of the labour force employed either directly by the State, local government or public corporations and the amount of money involved.

Fourth, the close interdependence of national economies and the measures necessary to be taken to protect internal finances and to secure satisfactory balances of payments, in a financial world situation where each nation's economy is vulnerable to the upsets that occur in others and where internal financial stability depends upon so many sensitive external factors.

This is the age of domination by money in the world of capitalism. The common factor introduced as the medium of exchange through which the value of each commodity could be expressed in exchange and distribution has become the master controlling every aspect of economic and human activity in the process of producing and distributing the things needed to sustain life.

It seems to me for these reasons central government has to engage more and more in planning, leaving less and less to the free play of markets and unplanned private enterprise operations. The essence, or content, in social terms have been described as 'socialization'. The change is away from uncontrolled operations by a legion of contending concerns, towards fewer and more concentrated undertakings, having to operate under controls and directions that emanate from a central body, these bodies being the agencies and departments of central government, the regional planning boards and through financial and credit policies. The State is now compelled to oversee in an ever increasing way the economic and financial activities of the nation.

One can readily appreciate the arguments that can be put against this general thesis. The most obvious one will be that it is not possible to 'plan capitalism'. Capitalism is essentially a social and economic system based on private enterprise, of production primarily for sale and not for use, with a system of distribution, because of its private enterprise nature, being based on competition. Because of these inherent anarchic characteristics, planning is inhibited and is a contradiction of everything that capitalism means and stands for. The Conservative case against nationalization of industry is that it removes competition from within the industry and the removal of the striving for maximum profit in a competitive industry takes away the dynamic to economic growth and efficiency. In other words planning is the negation of competition and a strait-jacket for the profit motive.

For these reasons it follows that effective planning is impracticable without the extension of public ownership. Planning in a static society, which aims to achieve a predetermined rate of economic growth, to preserve full employment, extend the social services and at the same time preserve the *status quo* in capitalism, is as feasible as reversing the law of gravity so that things that are down go up. Planning presupposes the existence of a central body with some power of direction and supervision, and would require subsidiary planning bodies over groups of industries. This central authority could only be that of the Government.

The term 'socialization' has been used to describe the general trend both in the character of ownership and the economic forms of organization, together with the more direct and greater involvement of the State. It is a term that should not to be confused with socialism. Socialization applies to 'form' whereas socialism applies to 'content'. The form of ownership and the form of direction and supervision has a more social form than the narrow private one of the past, but it is not used to suggest any fundamental social change.

This process of what we describe as socialization has not been a development resulting from some conscious purpose, but is a product of social and economic necessity and has already made fairly big inroads into our society. Events which have the force of revolutionizing modes of production, causing huge social problems as the changes take place, will compel a continuation of this development and bring central government more and more into direct social and economic planning activities. This is the only way that the changes taking place can be accommodated and used to the advantage of the people as a whole and in the interests of social progress. This is a positive process of change producing a new relationship in the balance of power as between socialism and capitalism in this country favouring the forces for socialism. The greater the power of local and central government and of the public corporations within the economic structure of the country and the larger the role of the Government in social and economic direction and supervision, the greater the power and control exercised by the trade union and labour movement – the greater the changes in the form; the more compelling will become the need and urgency for changing the content of social relationships. It is idle to believe that this whole complex of change can have no impact on the strategy and tactics of the trade union movement. The impact is of great magnitude and although some adjustment of trade union strategy to meet it is attempted, it is limited and slow – and conditioned to the theory of the 'inevitability of gradualness'.

The discussion on how this strategy ought to change is left to a later part of this study. Indeed, the central purpose of the study is to review trade union strategy and its structure against the background of the changes that have so far been surveyed.

THE CHANGED STATUS OF TRADE UNIONS

Compared with the situation between the wars, the status and scope of the trade unions' function and influence has changed dramatically. Between the wars, as previously stated, the situation in industry was one of continuous strife and conflict. The employers were arrogant and arbitrary in their treatment and attitude to trade unionism, the initiative being mainly with them in atttacks on wages and conditions of employment. The relationship with the State was very little better, with the unions usually in the role of the supplicant, seeking Government aid to achieve much needed reforms in industrial and social conditions. Although there was a kind of uneasy peace following the Mond-Turner talks and the collaboration which was developed after the 1926 strike, it was largely confined to the upper crust of industrial relations, having little effect on the character of collective bargaining or the results that emerged from it.

The emergency of war in 1939 brought this period of open conflict to an end. The nature of the war itself and the fact that the labour and trade union movement fully supported its aims produced a new situation. Hitler, in his attempt to conquer the world and impose a fascist dictatorship upon it, creating death, untold hardship and misery, also had the effect of uniting the peoples in the countries attacked in defence of democracy and liberty, at least for the limited period of the war, and for its vital, though limited, aims. It was during the war and especially following the establishment of a united war cabinet, which included leaders of the labour movement, that this new status was born. It is probably true to say that it was the fight for national survival against fascism that compelled the traditional, diehard, conservative rulers and the captains of industry to acknowledge in practical form the importance of trade unions. The enormous tasks facing industry to sustain and increase the war effort required the maximum effort from the workers and the removal of all obstacles to increased production, and to achieve this, the co-operation of the trade unions was vital. I well remember many pits during this period working a seven day coal-production week to help the war effort and this kind of gesture was taking place throughout industry. It was during this period of national extremity that the roots of many new forms of union activity were put down from which new industrial organizations grew and new forms of consultation between unions and employers, and unions and the Government, were accepted as necessary and useful.

To put the change in trade union status in more practical terms, some record of the actual situation is worthwhile. In industry there are now fewer areas where employers refuse recognition and where these do exist, new legislation supporting the principle of union recognition will make it very difficult for employers to keep up their resistance. In fact, the general attitude of employers and management is that they regard trade unions as a necessary and advantageous entity in the field of collective bargaining and in other relations within industry to facilitate discussion of affairs that affect the general body of employees. It is now generally acknowleged that without such representative organization the situation within industry would be much worse than it already is, with a far more fragmented state of wage arrangements, far greater disunity among the employees with a greater likelihood of disputes and strikes.

Few employers will now introduce a new system of production or a new mechanical device affecting the production process without first consulting with the leading union representative at the plant. In place of the former arbitrary action where the manager or the employer considered he had the divine right to do as he liked in his pit or factory and that workers were not entitled to any voice in the changes that could, and usually do, drastically affect their working lives, there is now consultation and, invariably, the assent of workpeople is sought. This does not mean, of course, that every new device or system of work is not proceeded with if the workers object, but it does mean that their viewpoint is put to management and in most cases considered, with the probability that their objections are answered and further clarification provided by reasoned explanation and often experimentation and visits to centres where the new device or system operates.

It would be stupid to suggest that this is a universal practice, but it is mainly true of the plants and industries where trade unions are strong and well established.

It has been said before that the aims and objects of the trade union movement remain constant, although the environment in which it has to pursue them is constantly changing. The 'protection and advance' of the interests of the workers is a much more complex undertaking in the second half of the twentieth century than it was in the first. Of necessity, it involves a much wider participation in the general process of industrial, social and political life and administration of the country. The range of trade union organization has widened, too, and now includes in its membership all classes of employees. To the early craft and manual grades have been added the clerical, administrative and supervisory staffs up to and, in some cases, the lower levels of management and these spread right across industry and commerce. This extension in organizational scope is of vital importance in view of the more recent and rapid growth of a huge superstructure on modern economies. The proportion of the labour force employed in the physical work of distribution and production gets relatively smaller while the services erected above these two main economic operations gets bigger. A large part of the increase in this type of labour is socially unnecessary even in a capitalist system operating normally, but in a situation of intense market competition the advertising and selling agencies are increased, involving inflated activities in a number of related industries and services. A larger part would be socially unnecessary in a properly organized society since they are a parasitic growth to this form of society, such as middlemen, the vast mass of insurance operations in existence to profit on the needs of people to pay for protection in sickness, times of adversity and interment at death – to say nothing of the army of lawyers and their staffs engaged in the simple though for them highly profitable operation of conveyancing property. These are but a few examples of the build up of a huge superstructure which contributes little to the main functions of the economy but are a direct factor in the causation of inflation, since they form a significant part of consumer spending, but add little to the wealth that is created.

This development is a by-product of the giant organizations and monopoly organizations now shaping the structure of the economy; of the extension of the application of science and technology to industry and the growth of research organization that it requires and the growing involvement of the State in social and economic operations, creates an evergrowing army of what are now called white-collar workers who have the potentiality in numbers of being the largest section of employees by occupation in trade unions.

During the war years, too, joint committees were set up

at production level to maximize output and these committees were representative of management and the main organizations in the undertaking, and operated in industry generally throughout the war. From this experience the Nationalization Acts which followed the conclusion of the war had written into them the obligation by the Boards appointed to run the industries to establish formal consultative machinery.

In the Coal Mines Nationalization Act the obligation is part of a two-barrelled duty imposed on the National Coal Board to 'consult' with appropriate organizations in the industry to establish joint machinery for the terms and conditions of employment; and consultation on:

'. . . questions relating to the safety and health or welfare of such persons; the organization and conduct of operations in which such persons are employed and other matters of mutual interest to the Board and such persons arising out of the exercise and performance by the Board of their functions.'

This became the pattern for nationalized industries and provided the employees in these industries through their unions, with a voice in the running of the industry. This, of course, does not mean participation in management decisions. It is not in any sense workers' control of the industry, but it does give the worker and the union the right to criticize and to put forward proposals and alternatives to management proposals affecting the operations of the industry.

Although it is true to state, as has been done above, that the unions at production level do not participate in management policy decisions, there can be matters where joint decisions are sometimes taken. In situations where absenteeism is an acute problem affecting productivity, in a number of places in the coalmining industry the habitual offender is interviewed by a small committeee of management and union officers; they might issue a warning if there is no improvement and if a second interview takes place, still with no improvement in attendance, the committee may decide to demote to a lower paid and less important job, and, in very bad cases, to dismiss.

The actual implementation of such a decision is the managers, but in such circumstances the union would not provide assistance to the member should he wish to contest the decision. This is an example of joint decisions in a very limited field and where such involvement by the union is only in the most chronic cases and after intense effort to get an improvement in the workman's attitude to his job.

The structure of the consultative machine in nationalized industries has its basis at the point of production and at the decisive control and policy making levels in the organizational structure of the Boards. In the coal industry this meant four levels until the recent reorganizaton which eliminated the divisions, leaving the pit, area or region, and national centres at which consultative committees operate. At pit or plant level the issues upon which consultations take place are, naturally, the intimate problems of production units such as quantity and quality of output; labour supply; materials supply, upsets and dislocation and management policy toward them; and, usually from the management side, poor attendance, labour problems and other deficiencies of the workers. The union representatives participate in the matters raised by management and draw attention to deficiencies of organization, bottlenecks, poor quality of supplies and so on. On every agenda in all the committees in the coal industry safety, health and welfare are standard items discussed at each meeting. At this lower level there is argument, often vigorous and aggressive, although rarely creating serious loss of goodwill, but out of it the unions know the plans and problems of the management, with management being made aware of the views of the workers and the unions.

At area and national level the committees may discuss similar problems but in the context of broader considerations. At area or regional level the committee becomes representative of a number of production units, and national level takes in representatives of the executive committees of the unions operating in the industry. In both these committees the discussions are more general in character and, in addition to the internal problems of the industry, are concerned in marketing and financial conditions and at national level with Government economic policy as it applies to the particular industry.

This is the formal process of consultation, but probably more important is the informal and intimate consultation that has grown up at all levels of management. The extent of this and the relations between the union and management is probably at its most intimate in industries experiencing great difficulties and where serious problems of a social and human character arise. In any case, this is the situation in the mining industry. The informal consultation is close and personal, making possible the passing of information and explanation in greater detail and on aspects of problems that would not be possible or desirable in the more formal meetings. The traffic in this more intimate relationship is two way and gives the manager and the union official a better understanding of their mutual difficulties and attitudes.

The consultations and discussions take place on the basis of mutual confidence and is a well-developed procedure in the mining industry at all levels which makes it possible for both sides to anticipate events and problems and to budget for them in advance.

Arguments are levelled against too close a relationship between unions and management, but, provided the union retains its independence and the union representatives their integrity, it is to the advantage of the work-people.

This more informal relationship is especially important during periods when reorganization and technological innovation is being undertaken. The effects flowing from such change create problems that are not normal to the routine matters arising from the day to day operations. This is especially true where automation is introduced, where even the 'brains' function of labour is being taken over by computers. The problems encountered here are new and difficult to anticipate and provide for, and daily intimate contact on the spot is indispensable in such situations.

In private industry the system of consultation is, in the first place, not obligatory on the employer. Where it exists, it is by voluntary agreement between employer and the unions; in a few industries it follows the organizational form operating in nationalized industries. Generally, however, workers or unions have not the same interest in engaging in this activity as they have in nationalized industry. More general is the informal consultation which is less constant than in the public sector and is mostly related to special situations as they arise.

However, the general position throughout industry is a big advance on the state of affairs between the wars and the significant and total change is worker participation in what previously was regarded as the exclusive domain of management and employers.

It represents, too, a big advance in the accountability of private enterprise firms to their employees in their methods of running industry. They are also accountable to the Government and their operations as well as their attitudes to labour can be brought before the bar of public opinion with resulting pressure upon them to change their policies where this may be necessary. Any developments considered to be anti-social can bring them under the scrutiny of the Monopolies Commission upon whose judgement the Government can take action affecting their future operations; if their undertakings are inefficient, proposals can be made for measures to improve efficiency by the Industrial Reorganization Corporation and pressures exerted for the implementation of its recommendations. More and more, through attention by local government, by the regional planning organizations and by the interest taken by Members of Parliament, the spotlight is focused on deficiencies, bad labour relations, or poor performance and results, again producing pressures for change. In other words, privately owned industry cannot so easily be run haphazardly or in the arbitrary way which was too often the case years ago. Employers and management are accountable publicly for their stewardship, both in respect of their performance in production and in their attitude and behaviour to their employees.

What has been argued in connection with the changes in the responsibility of employers to the nation and the status the unions have achieved in relation to the employers is equally true of the trade union movement as a whole in relation to central government. Central government, no matter its political complexion, must now consult with the trade union movement, that is, the Trades Union Congress and, when occasion demands, with individual unions, on any major question affecting the economic, social and political affairs of the country, which in other words means the general conditions affecting all members' and workers' conditions of employment and living conditions, or the problems of particular industries and communities.

This possibility of meetings between individual unions and Government Ministers or even the Prime Minister is a fairly regular procedure particularly in industries going through a period of special difficulty. Over the past decade meetings with the respective Ministers, and occasionally the Prime Minister, has been a regular occurrence, where discussions centred on the run-down of the industry and its effects on job security in the industry, and employment opportunity in the badly affected areas.

The Annual Reports of the General Council provide an insight into the numerous matters that fall for discussion between the Council and various Government departments. The subjects dealt with cover almost every aspect of life inside this country. In all of the discussions, the Council is in the role of advocate for changes, for improvement in this or that provision, or, as we have already intimated, for major changes in Government policies affecting the economic and financial state of the nation. This aspect of its work is now a main activity for which it has equipped itself with such an expert staff of research workers and advisors that its case to the Government on these important matters is as authoritative as that of the Government itself.

In the operations of central government, the trade union

movement in Britain is an established and powerful institution that must be consulted and whose views must be taken seriously into account by the Government in the formulation of its policies.

Undoubtedly, the greatest advance made in the direct involvement of the trade unions with Government policy is the recent development where each year the General Council produces its own assessment of the state of the nation, its own proposals as to what level growth is possible for the year ahead with the measures which could be taken to achieve this, and the overall average rate of wages increase that are practicable within this economic and financial policy. This assessment is presented to a conference of executives at the beginning of each year and – although it is a take it or leave it presentation, a vote to accept or reject the report as it stands – it does provide a very strong platform for the General Council in its discussions with the Government. This is a form of direct consultation additional to the discussion which proceeds within the broader National Economic Development Council.

There has never been a time in the history of the British trade union movement when it has been so powerful in the Council of the Nation, or where it has had the opportunity to exert such an influence on the policies pursued by central government.

The important aspect of this new role is the vital need for the trade unions to maintain their independence as a policy making body and their freedom to pursue causes by the means they consider most appropriate. Without this the present loose forms of consultation, the joint organizations comprising Government, employers and unions, could grow into an integrated corporate organization issuing agreed policy decisions to which the trade union movement would be tied. This is why it is so important in the present complex situation that democracy is extended to ensure greater participation by men on the job in policy making within the unions, and industry.

There is no identity of interest between employers and

work-people and trade unions must retain their freedom and independence to oppose policies they disagree with by whatever means they may decide. This is fundamental to the purpose of trade union organization and must be preserved. This feature of trade union status and representation in councils and committees set up by the Government is not only at national level. It is a development down through the whole administrative organization of the country. Trade unions in the regions have representation on a whole variety of joint bodies dealing with different features of the economic and social life of the regions. Although the bodies are usually only advisory in character, they represent an important forum where the views of workers and unions can be pressed. The changes are general and represent a changed environment in which unions have to operate. Trade unions are no longer organizations whose interests and functions are limited to wages and conditions in industry; they are part of the whole economic and social fabric of the nation.

Anyone who has come up through the trade union movement as I have, and has experienced the turbulence and hardship of the inter-war years, cannot fail to be impressed by the changes that have taken place. They are changes that have not arrived spontaneously but represent generations of continuous struggle for improvements. The contrast between then and now is tremendous, but this is not to say it is all that it ought to be. But the contrast cannot be denied, for the improvements are substantial and wide ranging. The trade union movement is accepted universally as a responsible institution that has to be reckoned with by both the employers and the State. Its new status has created the need for the reorganization and strengthening of the power centres of the movement, centres capable of meeting the new combinations dominating industry and, nationally, in relation to the more direct role of the Government. Changes in this direction are already proceeding and Congress and the General Council are being given more power to act for the movement as a whole and over a bigger area of questions. This trend needs continuing and accelerating if the unions are to effectively represent their purpose in these new conditions.

Up to now, this study has been concerned with contrasting the inter-war years and the post-war period, without drawing too many conclusions. The period covers fifty years and what has been presented is but a brief sketch of events and their background in both periods. The emphasis has been on aspects that have a bearing on the subject matter being studied, trade union structure and strategy. There have been changes and modifications in both and the question posed is whether further change and modification is necessary and possible.

The main changes in the contrast can be summarized into some general conclusions.

First, the structure of the economy has radically altered in the post-war period compared with the inter-war years. A much larger sector of the economy has been withdrawn from traditional capitalist or private enterprise control and direction. The size of the investment and expenditure, the labour force employed and economic influence exercised, is enormous. This with the direct financial and economic activities of central and local Government is a formidable power in the relations in industry.

Second, allied to this development is the creation of huge organizations in industry and commerce, referred to by economists as 'conglomerates', but simply put, represent giant monopolies of great power operating mostly on modern techniques and backed by huge capital resources. Foreign capital is in control of many of these giants and their enterprises know no national frontiers. The trend is for these giants to get bigger and more powerful.

Third, the character of the 'employer' has dissolved into an intangible, unidentifiable mass, relegating to managerial specialists the job of running their industry, with interests and enterprises operating in more than one country.

Fourth, a technological revolution based on applied science and automation is under way, disturbing job relations and opportunities in industry, requiring massive capital support and creating social dislocation, compelling central government to take action in many measures involving substantial capital expenditure.

Fifth, central government has moved into industrial relations and wage determination and, judged against the economic and financial state of the nation and its relations with the outside world, this appears to be a permanent sphere for Government intervention and supervision, whether supported by legal sanctions or not. In addition, the condition of balances of payments, the continuous climate of inflationary swells and the commitments entered to the World Bank and other sources from which the Government has borrowed huge loans, and may have to borrow more, together with heavy internal expenditure within the country on services and development, are supporting factors justifying the assumption that central government for the foreseeable future will maintain its policy of influencing all forms of incomes and, wherever it is able, of expenditure too.

Sixth, money wages are not being attacked directly by the employers as in the inter-war years, but their purchasing power is being continually eroded by rising prices. This, plus the measure to apply new techniques and rationalization to increase productivity will create constant pressure for wage increases. The demand for wage and other improvements are, in present circumstances, not just questions for discussion with representatives of employers, but must inevitably involve central government. Wage conflicts in the future will be conflicts, in the main, involving the Government. Giant economic organization plus central government is the setting confronting unions now and moreso in the future.

Seventh, the trade union movement is now a powerful institution in the nation's affairs and in industry. It is able to exercise influence in most spheres of economic life. Individual unions in industry confront much larger economic organization against which their strength has to be measured and organizational structure is an important factor in their unity and strength.

Eighth, the Trades Union Congress and General Council

are required more than ever to speak and act for the trade union movement as a whole and with the centre of gravity in industrial affairs being the Government, this responsible function is bound to increase. This raises two main questions affecting the General Council. One, as to whether under the Constitution its policy declarations have any force in the trade unions affiliated, and whether the present form of election to the General Council and the principle upon which it is based provide the most representative leadership covering industry and membership.

Compared with the position in the inter-war years, the changes are such as to bring into question whether the trade unions have adequately assessed the significance of the new alignment of forces and have adapted union structure, services, strategy and tactics to effectively meet them. The summary of change set out above is the new environment in which unionism has to operate and it is in relation to this that structure and strategy is being tested.

THEIR PURPOSE IN NEW ENVIRONMENT

It has been asserted previously that the primary purpose of the trade unions is that of 'maintaining and improving' the wages and working conditions of its members. This is their economic purpose, but its pursuit in the conditions of modern society becomes more and more political in character. Trade union purpose has therefore a much wider definition and application now, being concerned with the health, welfare, education and living standards of those for whom it is in existence to serve from their cradle to the grave. It is a concern expressed in continuous pressure and representation for economic and social reform, requiring more and more legislative and other forms of political action for their realization. Thus the prosecution of purpose in this changing environment is economic, social and political, with the emphasis upon influencing ever increasingly direction and control in the determination of policy in these areas of activity.

If the assessment, made earlier in this study, of the changes that have taken place are roughly accurate, the inevitable conclusion must be that individual unions are relatively less powerful now than they were before these changes assumed a decisive influence in industry and the economy generally. If it is a correct conclusion that the purpose of maintaining or improving the working and living conditions of work-people is now no longer capable of being freely determined by unions and employers, but is subject to external influences which neither can neutralize or effectively control, then it seems obvious that the power to determine wages and living standards by agreements made within industry has been substantially restricted.

This, it seems to me, is a fact of life that has to be faced. A particular union may be more united, with increased membership and more militant in outlook, stronger in itself but still weaker in relation to the combination of power represented by the giant amalgamations or conglomerates and the State.

There has been a change in the balance of power as between unions and employers in industry and it is one that substantially accentuates the weakness of the trade unions caused by their existing fragmented structure. Multi-unionism, the existence of a number of unions operating within industry, within particular employing companies and in the same workshop intensifies this power relationship to the disadvantage of the trade unions.

It will be argued, no doubt, that full employment and the relative scarcity of certain types of skilled labour is a compensating factor reducing the adverse effects of this change in the balance of power. It is, of course, true that when the demand for labour is greater than the supply, considerable bargaining advantage is gained by the trade union operating in the area where this condition applies. Full employment, however, is by no means a universal bargaining advantage for all unions and in all areas of the country. There are contracting industries, there are areas of heavy unemployment, industries where modernization will create labour surpluses, all imposing serious limitations upon the ability of the unions faced with these conditions. Exploiting the advantage of full employment or scarcity in particular geographical areas and certain occupations and industries leads to intensification of earning differentials within national industries and often produces disunity within unions. The belief that a union because it is big or in a key industry is able to succeed by 'going it alone' may be given more credence than it deserves because the consequences of going it alone cannot be confined and can have a chain effect on workers over a wide area of industry.

It is my firm conviction that the success of individual unions now and in the future will depend more and more upon the influence the trade union movement as whole is able to exert on the nation's economic and financial policy. It is the general economic and financial climate of the country that will condition the outcome of union bargaining, and the creation of a favourable climate therefore becomes a major purpose for the trade union movement.

This conviction is not based on an assessment of the policy being operated by a particular government. It is based on a development of modern capitalism towards ever greater control by central government over the internal finances of the country, including income levels, a development that is not peculiar to Britain, but is a general development in all countries.

The character of the Government's financial policy as we have so far experienced it, intrudes into the bargaining procedures in industry. It is a reasonable conclusion that even if this form of intrusion ceased to operate, some alternative form, perhaps less rigid and direct, would almost certainly take its place. So long as it is considered necessary to limit consumer spending (and this is likely to be a continuous requirement in modern conditions) some form of central control will operate. This is not intended as a justification for such controls but it is an acknowledgement that in the conditions of modern capitalism incomes control is now as much an integral part of a country's monetary policy as direct or indirect taxation.

It goes without saying, of course, that if central government changed its policy in relation to world affairs, or to the wealth of the rich, or defence expenditure, or the export of capital, then the extent of its intrusion into the field of wages determination could be less. While these are aspects of policy that are legitimately agitated for, it also means that until such policy changes are realized controls will continue in one form or another. Listening to and reading some speeches of politicians and union leaders one gets the impression that Government intervention in the field of industrial relations is inspired by some malevolent antipathy towards trade unions and workpeople.

The policy may be wrong and deservedly condemned, but it is as well for the purpose of changing it that it should be understood for what it really is, which is a product of existing economic and financial conditions. Otherwise, people may be deluded into thinking that a change of Government will bring it to an end, whereas, as is surely the certain truth, a radical change in policy towards the social system that creates it is the only way.

If this judgement of the effect of the change in the balance of power as between unions and employers is accepted, then it follows logically that in these circumstances the central organization of the trade unions must assume a far more decisive role in the prosecution of trade union purpose. It has already been stated that the trade union movement has secured for itself a status and influence in national affairs of some significance in the post war history of this country. It is of course always open to argument that it could have been more effective had it been more vigorous, more militant, more resolute and decisive in using its economic power. But judging it by its own constitution, by the composition of its leadership, and the limited power conceded to it by the affiliated unions there are few genuine grounds for expecting anything very different from what has in fact taken place.

However, the burden of my argument is that now and in the future it is the power and influence of the whole trade union movement in fashioning national policy that will be of decisive importance for the individual unions and their members, and this must inevitably mean more power and authority to the Trades Union Congress. The prerequisite to any reorganization and strengthening of the TUC for undertaking the role that history is placing upon it is the restructuring of the trade unions that make up Congress. But frustrating the full realization of this role is the continued acceptance and, even worse, the defence of the chaotic multi-union character of trade union organization in industry in this country. Some union leaders can boast of assisting in the setting up of the sixteen industrial unions in West Germany after the war, and rightly acclaim the structure they created as a model one. But when they face the proposition that a move towards industrial unions should be proceeded with in Britain, they oppose it as 'undesirable.'

3

Trade Union Structure

THERE have been many discussions in the British trade union movement over the past half century, centred on trade union structure.* Invariably, the principle of industrial unionism has been the form advanced by those who advocated structural change. The general concept of industrial unionism is that unions should be based on industry taking into membership all workers employed in that industry irrespective of grade or specialized occupations.

The Webbs, in their *History of British Trade Unionism* (p. 548), describe the concept as 'an organization based on the whole of an industry such as engineering, housebuilding, mining or the railway service, in which all the operating crafts and grades of workers would be associated in a single industrial union; in contrast with the earlier conception of the separate organization of each craft throughout the whole kingdom; such as that of carpenters, the enginemen, the engineering mechanics, the clerks, and by analogy the general labourers, in whatsoever industry they may be working'.

They point out, too, that this concept of union organization was not only advocated as a change in union structure but as the basis of a new social order. The agitation for industrial unionism was strengthened by both the Guild Socialists and the Syndicalists. The Guild Socialists proposed a social structure incorporating the State, the trade unions and consumers, as the basis for running industry, the social services and the

* As a result of further resolutions at the 1969 Portsmouth Congress the question is again under active review. The resolutions were promoted by the smaller unions and it is of some significance that the big unions took no part in the debate.

affairs of the nation generally. The aim of the Syndicalists was to construct unions in each industry as a means of improving workers' wages and conditions in the short term and of ultimately taking over complete control of each industry as the basis of a new social order replacing capitalism.

The 'Miners' Next Step' produced by a group of South Wales Miners leaders in 1912, after stating the immediate objectives for organization and for wages and hours put its ultimate aim in the following terms:

'That our objective be, to build up an organization that will ultimately take over the mining industry, and carry it on in the interests of the workers.'

The political essence of the theory is that the strength of the working class in their struggle against capitalism is in their industrial organizations, and industrial unions would be the vehicle through which fundamental social and political change would be achieved. It rejected parliament and political parties as the means by which workers' emancipation could be secured, contending that real power was concentrated in the economic organizations of the workers.

Indeed, this outlook is still strongly held in some trade union regions. During the inter-war years in the western part of the South Wales coalfields this conception of the syndicalist role of the trade unions was still being actively discussed at miners' evening classes. This belief that economic struggle is still the only effective form of struggle to achieve reforms and radical social change is in my view still a powerful influence in the trade union movement.

However, the discussion in Britain over the last forty years on the need to restructure the trade unions has not projected industrial unionism as the vehicle for revolutionary change. Those who have argued the case for industrial unionism have concentrated on the advance it would represent in the effectiveness and efficiency of the unions in the pursuit of their more limited aims. A study of the debates that have taken place in the Trade Union Congress on this proposition reveals very little change in the arguments for and against. The two main periods of debate were first between 1924 and 1927 and more recently between 1962 and 1964.

THE 1925–1927 CONGRESS DISCUSSION

It is perhaps worthwhile, in view of the importance of this subject now, to summarize a little of this history if only to expose how superficial the attempts at restructuring have been. It is important, too, that the arguments for and against should be identified and analysed.

At the Scarborough Congress in 1924 the Miners' Federation moved a composite resolution which instructed the General Council to prepare a scheme to reduce the number of unions, for organization by industry and for unity of the trade union movement.

The central theme of the resolution and the debate was Industrial Unionism. The resolution was seconded by the Amalgamated Society of Locomotivemen and Firemen. The General Council reported the outcome of their discussions and consultations with the unions to the 1927 Edinburgh Congress, the report being prepared and presented by Walter Citrine, the then General Secretary of Congress. The report was based upon replies received to a questionnaire to the affiliated unions as well as verbal evidence from a number of them. The report divided the type of existing unions into five categories; craft; industrial; occupational; employmental and general workers.

The report revealed that some craft unions were in favour of organization by industry, but they defined industry as their own particular craft. The unions it seemed by their replies defined the boundaries of industry as being determined by three principal factors:

(1) The commodity produced or service rendered.
(2) The tool operated.
(3) The employer or group of employers.

In commenting upon these definitions the report stated;

'Those contending that the commodity determines the boundary of industry would argue that those engaged in the production of cotton materials are in one industry, those producing woollen and worsted materials in another. If, however, the tool is to be the determining factor the engineers and maintenance men in both industries might claim they belonged to the engineering industry and not to either the woollen or cotton Industry.'

It would not be unreasonable to have expected that the General Council would have anticipated that individual unions would approach the proposition for industrial unions with their primary concern concentrated on their own self preservation as independent units, and would have prepared answers to such attitudes. The operative part of the resolution instructed the General Council to prepare a 'scheme', and this should have been done at least as the basis for discussion. Instead, however, the General Council's verdict was that 'after very careful consideration of the problem "they" had been forced to the conclusion that it is impossible to define any fixed boundaries of industry, it is impracticable to formulate a scheme of organization by industry that can be made applicable to all industries'.

In moving the reference back, Mr. Jack Tanner of the AEU acknowledged that the job was difficult but not impossible. No one expected clear-cut boundaries and there were bound to be what he described as 'ragged edges'. He suggested using the group basis of election to the General Council as a possible definition of industry for the purpose required. In his view the General Council should have presented a scheme for discussion.

The main speech against the reference back was made by Ernest Bevin, of the T & GWU who referred to his own union including within its organization 'gravediggers' and 'midwives'. He argued that members could not be forced to transfer from one union to another against their will.

'You have got to study human psychology', he said, 'and the most conservative man in the world is the British trade unionist when you want to change him.'

The reference back was narrowly defeated, the vote being 1,809,000 for, and 2,062,000 against, a majority of only 253,000.

The policy emerging from this discussion directed itself to increasing the tempo of amalgamation and federation as the means of reducing the number of unions and easing the competition between them. But history was to show that amalgamation was a slow and painful process, even for the miners, the promoters of the original resolution. They continued as a federation of autonomous districts until 1945, twenty years later. In fact, many of the districts were themselves made up of autonomous regional unions which in some instances were not united until the late 1930s.

The question was again raised in Congress in the 1940's and resulted in a very detailed report being prepared in 1944, by Vincent Tewson, Congress Secretary. However, the outcome from the standpoint of any change in policy was negative. For some reason this report seems to have been ignored in the later discussions.

1962–1964 CONGRESSES – STRUCTURE AGAIN

It was the Post Office workers union who tabled a resolution for the 1962 Congress which again focused discussion on 'structure'. The resolution called for an examination of the possibility of reorganizing the structure of the TUC and the trade union movement. No design for restructuring was suggested in either the resolution or the movers speech, the plea being that the General Council should 'examine and report'.

The most forceful speech was made by the late Sidney Hill, who seconded the resolution for the National Union of Public Employees. He was critical of the General Council for their attitude to this and other matters and argued that it was necessary to reduce the number of unions in order to cut out overlapping and duplication. The issue of industrial unionism was not mentioned in the debate and it would appear that it subesquently became the centre of the examination by the General Council on the initiative of George Woodcock, Congress General Secretary. In his report to the 1964 Congress he stated, 'I was the one who put forward the idea of industrial unionism as a proposition' (page 375 of report).

The resolution was carried unanimously after Congress was informed that the General Council was in support. The issue to be defined first was that of purpose, contended the General Secretary. 'Structure is a function of purpose' was his substitution of Walter Citrines earlier formula, 'Function must determine purpose'. First we must determine our purpose, 'what we are here for, and then we can talk about the structure needed to do what we are here for'. This was the advice given to Congress on behalf of the General Council. Then only four years away from the celebration of its hundredth anniversary one could have assumed that the purpose of the trade union movement was self evident from the growth of its interests and the range of its activities.

In his autobiography, Walter Citrine, commenting on his memorandum on structure to the 1927 Congress outlines his view of the relationship between function and purpose.

'I pointed out,' he writes, 'that the Trade Union Congress was itself a Federal organization, I emphasised that it could not lay down any plans for reorganization which its affiliated unions could be compelled to accept. No matter what the TUC said, the individual unions always had the last word. . . .' 'I set out to examine the objects of Trade Unionism as stated in the rule books, but whilst the majority of unions had been established with the primary purpose of dealing with wages and working conditions, in only a small number of cases was there a clear recognition that the Trade Union movement had a wider purpose. "Function must determine purpose" I wrote, "and that the type of organization which will suit the minimum needs of a union's own membership will not necessarily be best for the attainment of all the broader objects". I defined these objects as (1) improvement in wages and conditions, (2) a measure of control of industry, and, (3) the abilitv to defend workers against any onslaught by capitalism. . . .' 'As to defects', he continues, 'in trade union structure, they had long been apparent. I set them out as (1) sectionalism, (2) competition for members, (3) unions offering different rates of contributions and benefits for apparently the same service, (4) demarcation of work and, (5) lack of co-ordinated policy.'

Generally stated, this presentation of trade union objects or purpose as outlined by Citrine are true of today but their realization is in vastly changed conditions. The earlier part of this study has tried to outline what these changes are and the new environment within which trade unions must now function. It is this radical change in trade union function within industry and the economy generally that emphasizes the importance of structure, and accentuates the defects which are as active and apparent now as when he set them down in 1927. It was the importance of this change in function that the General Council ignored in its examination of structure in 1962 to 1964. In fact this second review of structure was a rather shoddy re-hash of the report on the same subject almost thirty years before and added nothing to its conclusions or recommendations.

Indeed the interim report to the 1963 Congress made it clear that the General Council had already decided against adopting any moves towards orientating union organization nearer to an industrial structure, and would adopt instead the same policy conclusions as their predecessors did in 1927. More mergers, amalgamations and federations were suggested, even in the interim report, as the way forward, really making the 1964 report not only a foregone conclusion but superfluous. The 1927 effort was described as 'undoubtedly the most thoroughgoing report so far produced' and its conclusions as a 'sensible and inevitable reflection of the different circumstances of industries and work-people'.

However, the summing up of this second exercise is to let nature take its course.

'Rather than attempt to draw up a scheme for widespread structural reform the committee proposes', as the Report finally put it, 'to consider how best to stimulate and guide piecemeal and ad hoc development by which changes in the structure of unions have come about in the past.'

In the past any change in union organization has been without conscious purpose and this is to be the design for the future, except of course, the design of the big organizations swallowing the smaller ones and consolidating power in larger organization spread horizontally across industry. Fewer unions there may be, but the antagonisms could be more intense.

In both 1963 and 1964 the burden of the reports to Congress and the speeches of the General Secretary were to the effect that industrial unionism was both 'undesirable and impractical' in British conditions. The arguments in support of this conclusion were fully set out in the 1964 report. Put briefly, these were: (1) There was no consensus of opinion on suitable boundaries to define 'industry'. The evidence advanced in support of this was the conflicting definitions already apparent between the TUC, Ministry of Labour and the Confederation of British Industries. The basis of representation on the General Council was on eighteen industrial groups, the Ministry operated on twenty four, and the CBI on 53! (2) There was a common interest of skills and crafts spread over industry best accommodated in unions based on such common interests. (3) That with horizontal unions men who move from one industry to another can remain in the one union. (4) There were difficulties created by the disparities in Friendly Benefit arrangements. (5) Some unions held the belief that union structure should be based on occupations and not industry, and (6) that horizontal unions can accommodate vertical trade sections with a large measure of autonomy.

In the words of George Woodcock, the General Council had again reached the conclusion that industrial unionism was 'not a runner in this country, nor likely to be in the foreseeable future'. He said that 'without exception every trade union we have met resisted the idea of industrial trade unionism', but then later rather qualified this with the statement that 'nobody is against industrial trades unionism where it can be adopted and made to work'.

However, nowhere in the report and discussion is there evidence of a serious examination of industries where it might be adopted and made to work. The whole approach seems to have been that the conversion from the present form of structure to the one proposed would be a single total operation, and not that of a gradual piecemeal development spread over a period of years as part of a comprehensive long term plan. No one denies the practical difficulties but there are none that could not be overcome by men who have the conviction that the change is not only 'desirable' but necessary.

The arguments advanced in support of the General Council's report rejecting fundamental change are in my view superficial and subjective. They reflect the conservatism that is a characteristic of a movement where tradition is such an overbearing influence. The attempts made to further streamline the organization of the NUM exposed for me the deep suspicion and even hostility that proposals for structural change can create. But the factors that make such change imperative can convince reasonable men, if the case is properly explained and understood. The attempt was not made by the General Council, and they fell back on these weak excuses.

Take the argument that there are no definable boundaries to industry. There are a host of conditions where man has defined boundaries where they are not determined by their own character. Man has fixed the dates which sets the beginning and end of the seasons, and one does not need a long life to realize that they are arbitrary and often seemingly ill-defined. The boundaries of industry for the purpose of trade union organization would have to be arbitrarily fixed as has been done in the majority of other countries where industrial trade unions are established.

Then again there is the argument that skills and crafts have a common interest in organization covering all industry. The implication is that this is the present form of craft organization. It is anything but this. Craftsmen are organized in all the forms that make British unions the strangest in the world.

The so-called craft unions are no longer in character with their names and recruit non-craftsmen just as other unions recruit craftsmen. For the purposes of rational wage structures and maintaining equitable wage relationships within an industry, and above all for ensuring maximum unity in relation to an employer, the case is ten times stronger for craftsmen being in a union with the men they work with in an industry. The argument also fails to take account of the changes taking place in industry affecting skills and crafts. The character of skills is changing and being merged and integrated into the total industrial operation obscuring the demarcation that used to exist between the craftsman and the non-craftsman.

The most puny of all the arguments advanced is the one which seeks to justify the perpetuation of horizontal unions because they facilitate men moving from one industry to another without necessarily having to transfer their union membership. The system of inter-union transfers is so well established that difficulties from this source are extremely rare. In the last decade few industries have faced the run-down in manpower like the coal mining industry, with high voluntary leaving to jobs in other industries. I can recall no single instance where difficulty of inter-union transfer has arisen. Apart from this, even with horizontal unions, because of zones of influence and demarcation considerations, union transfers have to be made. This, together with the reference to differences in Benefit payments, really represent the scrapings of the barrel. Benefit payments by unions are really devices for recruiting and holding members against competition from other unions. They are no longer socially necessary since the State social security benefits have been improved and would not be necessary as a recruiting device by industrial unions.

The final argument on the list is that horizontal unions could have vertical trade sections built into them and, as one speaker in the debate said, this made such unions industrial unions. There might be some merit in this claim if the trade sections embraced all the crafts and grades in that industry, in which case if such sections existed in every industry there would only be one central union and no need for a Trade Union Congress! That, I suggest, is really the non-runner in the conditions of Britain. The facts are, however, that these trade sections represent only one fragment among many others in most industries, and are part of the problem of multi-unionism in those industries.

If there is no turn towards industrial unionism what is the prospect for trade union development in this country? In the 1968 Annual Report of the General Council the trend to amalgamation is shown in the information of mergers that have been realized or are in the pipeline. The mergers, in the main are into the already massive unions spread across industry and commerce and covering a membership that includes the range of grades, crafts and skills to be found and catered for in each of them. Fewer unions and bigger giants means no essential change in structure and not necessarily an easing of inter-union divisions. As John Hughes stated in Research Paper 5 to the Donovan Commission (pp. 24-5):

'Thus the continuing development of a number of large and relatively "open" Unions has not resolved, and appears unlikely to resolve, what might be called the central structural problem of British Unionism, multiplicity of unions and overlapping of unions in particular sectors, or occupations. Indeed, by their nature they are more likely to extend overlapping. Sectoral rationalization through amalgamation has eased the problem in some cases, but no general "tidying" is in sight.'

The potential of this pattern of growth creates its own limitations. It can grow too big for any central leadership to be able to give consistent and effective leadership. Confined to one industry it is possible for a big union to provide detailed and intimate leadership from a national centre down through a single administration, but where a big union covers many industries of contrasting types and problems, leadership from a central source tends to become diffused and indirect and less detailed and intimate. In such circumstances leadership is either authoritarian and undemocratic or loose and divided sectionally and consequently ineffective. We have examples of both in the trade union scene in this country.

Horizontal unionism is the framework for multi-unionism and the disadvantages and weaknesses of this system of organization are not only colossal but damaging to the interests of work-people. Take a modest example of an undertaking with ten independent unions operating within it. There will be ten different union administrations with some responsibility for a section of work-people in it; ten different union officers to be called upon to concern themselves in a matter that might be common to all employed in the undertaking. It means ten different union administrations dealing with the undertakings problems of safety, health and welfare, as well as accident claims and legal representation as and when occasion warrants. The system is damaging because it involves spending double or treble the amount of money from the collective contributions of the employees for a service made less efficient and effective by the extent of overlapping and duplication involved. It is obvious, too, that recruitment of non-unionists is made more difficult by the number of unions interested and more confusing for the non-member for the same reason.

What is even more damaging is that the employees in the undertaking are organizationally divided in relation to a single management or employer. To act together in any situation usually means unofficial organization and action. But most unions now acknowledge that one union in one undertaking is advantageous to representation and negotiation, and some of them testified to this before the Donovan Commission. The limitation to this process or reasoning is in the fact that they do not apply it to the whole of a company's undertakings or to an industry. Nevertheless it represents a beginning to the acceptance of the principle of single union representation at least at one level of operations. The process of reasoning by which the principle of one union one undertaking is justified is the same reasoning process which justifies industrial unionism.

THE ADVANTAGES OF INDUSTRIAL UNIONISM

The report of the General Council to the 1964 Congress concentrated on the arguments which supported the conclusions they had reached. In citing the advantages claimed for the concept of industrial unions they referred to them as 'beliefs'. The report stated that those who advocated and supported this form of organization believed that a union embracing all workpeople in an industry provided a common interest which would give greater strength in collective bargaining; sectional claims could be more easily harmonized; a coherent wages policy would be less difficult to achieve; problems of demarcation could be reduced and controlled; a unified policy toward industry could be realized as well as in respect of the economy generally; fewer full time staff would be required while duplication and overlapping of functions as well as inter-union conflict could be eliminated. This is indeed a formidable list of advantages, which, if they can be proved to be more substantial than just 'beliefs', should convince unprejudiced trade union leaders that industrial unionism is at least desirable. The rank and file members of unions whose interests are in results and service from a union would, I am sure, support such a change.

The comment of the General Council on this list of advantages was opened with an acknowledgment that at least at first sight the advantages seem valuable.

'At first sight this list of advantages contains so much of value to work-people that those who reject industrial unionism seem to stand in the way of progress. But this is not so.'

I am convinced that opposition to a development towards industrial unionism is opposition to progress and that the 'conservatism' that this opposition represents is not as Ernest Bevin described an attitude of trade union membership, but very definitely an attitude more appropriately applied to union leadership.

My judgement of industrial unionism does not derive from

reading a list of suggested advantages, but is based on forty years activity and service in an industrial union, and in an industry that has thrown up over that period an assortment of problems and crises unequalled in any other British industry. It is important therefore that some study of the one union which is the nearest approach to an industrial union in this country should be made as part of this study.

In doing this I must first deal briefly with a question that appears inevitable by anyone reading the statement just made. Why were the arguments now being advanced in support of industrial unionism and based on the experience of the miners unions not put in the 1964 Congress? They were not made because, as with many other issues which do not directly affect a particular union, the proposal was relegated as one concerning other unions and not directly the concern of the miners' union. The attitude is grossly at fault, but is one unfortunately that is all too common with many of us.

It would be wrong to give the impression that the National Union of Mineworkers is a model industrial union. It is not and has defects and very definite 'rough edges'. The union grew out of the former Miners' Federation of Great Britain and has functioned in its new organizational form since 1945. The MFGB was a federation of autonomous district unons whereas the NUM is an integrated centrally directed national union. But the transformation to an industrial union is not complete. The NUM incorporates the former district unions as 'constituent areas' but left them with a measure of autonomy over their own funds, and some aspects of general policy, especially in the area of politics. Each of the constituent areas is registered as a union but written into their rules is the condition that they are subject to the over-riding authority of the National Rules. In a few of the areas there are separate organizations of craftsmen, but these are being progressively reduced by mergers into main constituent areas. Supervisors and clerks are organized in a special area of the national union and formed as a vertical unit within, but subject to the authority of the National union. Some clerks are in membership of another union who participate for this section only in a joint conciliation scheme. Colliery deputies and overmen are in a separate union affiliated to the TUC. In addition there are dual membership agreements with certain craft unions but the reason for such agreements are disappearing with the operation of national standards for craft apprenticeship and the existing facilities for inter-union transfers. It is my view that such arrangements are now superfluous and should be terminated.

Notwithstanding this somewhat incomplete industrial structure, the NUM has exclusive negotiating rights for all manual grades, including craftsmen, and is decisive in the negotiations for supervisors and clerks.

Generally speaking membership of the Union is restricted to men employed 'in or around' the coalmines which form the industry, and the principle agreements for wages, hours of work and other conditions of employment apply to workers within this definition.

In 1948, this limitation on the definition of 'in or around' was strengthened by a national agreement, called the Ancilliary Workers Agreement, which excluded all those employed 'off' colliery premises from the cover of mining agreements. This agreement took some time to implement, a deal of investigation, and sometimes arbitration to settle whether a particular job was on or off colliery premises. As an agent of the Union it was part of my job to deal with these 'on or off' problems, but after 1951 as President of the South Wales Area I had to deal with the consequences and transfer to other appropriate unions of those workers excluded by the agreement from the cover of mining agreements.

The workers employed on the coal-industries housing estates we transferred to the Building Trades Union; the wagon repair men were transferred to the AEU. We transferred to other unions all those workers for whom we could no longer negotiate terms of employment. However, it must be acknowledged that the agreement was applied less rigidly in some coalfields and this is a criticism of the National Coal Board, which allowed certain of their areas to default in implementing it.

This experience not only shows that it is possible to define and operate boundaries for industry, but that it is also possible to transfer bulk membership when the restructuring of a union requires it. If the will to make the change is present and careful steps are taken to explain the reasons for the change and to protect the special interests of those who may be transferred, then this experience shows that it is practicable.

My own view of the ideal industrial union structure would be to include within the union all grades, crafts and skills below the level of management, with clerks and other white-collar workers, supervisors, overmen and deputies and apprenticed craftsmen, in separate vertical sections within the union to meet their special occupational interests, but subject to the policy decisions and direction determined by the union as a whole.

Even with this less than ideal structure in the mining industry the union has not been troubled by demarcation disputes except for a few minor difficulties where separate craft organizations still operate. In all the areas where all grades and crafts are integrated, demarcation disputes never arise, and sectional interests are harmonized into the interests of the general body of work-people. What is infinitely more important for the effective functioning of a union in relation to an employer, is that this unity of interest gives the union far greater power and authority in representing and negotiating on behalf of this unified membership.

The administrative structure of the union is simplified and only a minimum of overlapping occurs in these few regions where these small and diminishing craft units operate. In general there is a single line of communication from the branch up through the area or coalfield office to the national centre and similarly from the centre to the branch. With an industrial structure the local branch is based on the colliery, the unit of operation, with the branch personnel who deal with the day to day problems, disputes, grievances and the hundred and one issues that can arise, being immediately and directly accountable to the local branch for their stewardship. Shop stewards as such are not to be found in the mining industry, because the branch is based on the pit and includes in its membership all manual and craft grades and the branch officers and committeemen undertake the duties allotted to shop stewards in other industries. Thus the organization dealing with the day to day industrial problems is an integrated part of the formal trade union structure, subject to its direction and discipline.

It may be argued, of course, that this type of structure and administration is possible in the mining industry because there is only one employer in the industry. This is naturally a favourable factor in many ways, its advantage being a single unified management control throughout the industry, a condition that could be extended to other industries with equal advantage.

But this favourable factor is present in many industries; in railways, electricity generating, gas, steel, post office, and air transport, in fact in all the activities and undertakings controlled directly or indirectly by the State. Their difference is, of course, in the presence of a number of unions in each of these industries. It is this factor and not that of ownership which militates against a unified trade union administration, which creates the conditions for discord, and dissipates the strength of the work-people in their pursuit of common aims.

INDUSTRIAL UNIONISM AND WAGES STRUCTURE

By far the greatest advantage gained from industrial unionism is the opportunity it creates for introducing and applying comprehensive and coherent bargaining procedures through which rational wage structures are possible. The experience of the mining industry in this respect is not limited to the period of nationalization. As previously mentioned in this study, a wages structure was introduced in the South Wales Coalfield in the 1930s by agreement with the coalowners, and the collective bargaining machinery still operating in the industry has its roots in the pre-nationalization period.

Because of the interest now shown in collective bargaining procedures following the report of the Donovan Commission, it will be of some interest to describe the procedure operating in the mining industry. The procedure is constructed in three tiers with a complete process of conciliation to finality in each tier. At the base there is the conciliation scheme for the settlement of disputes which arise from agreements or conditions restricted to a particular colliery. These are defined as 'pit questions'. The second tier covers questions that involve more than one pit or which arise from arrangements restricted to a district, and are defined as 'district questions'. In both schemes failure to reach settlement can mean reference to arbitration, which is inbuilt and where the decision from such arbitration is final and binding on the parties.

The third scheme covers questions and issues that are national in character, and since national agreements are now fast replacing pit and district agreements, this is the conciliation scheme through which most questions and disputes are dealt with. The procedure for initiating discussion on a dispute or grievance is the same in each scheme, starting at pit level between the man and his immediate overman or manager, and if unresolved referred to branch negotiators and through to the area union and if necessary to national level. The concentration is upon settlement without recourse to arbitration even on pit and district questions. Reference to arbitration of national questions is optional, and although there is always available a 'National Reference Tribunal' built in to the National Scheme it is very rarely called upon to arbitrate. The principle of compulsory arbitration was inbuilt into each tier, but was removed from the National Scheme on union insistence a few years ago. It is now open to the union to use methods other than arbitration to resolve unsettled National Questions if it so desires.

This is a comprehensive conciliation procedure which permits any question to be raised and discussed which arises from employment in the industry and, in practice, questions of general concern arising from a very broad definition of 'employment'. It would be misleading if the impression created by this description of bargaining procedure was of an industry free of friction and strife. My experience demonstrates that it is possible to have an ideal conciliation system and still be on top of the league for unofficial strikes and go-slows. The failure to reach a speedy settlement particularly of questions around which high feelings have been aroused can lead to strikes, and those who think that there is some magic way of reaching speedy decisions must live in a dream world. It is possible to get quick settlements if the union accepts the position of the employer whether it is satisfactory or not. But this of course is the negation of what is meant by 'conciliation'. There is no quick and easy solution to many of the issues that arise between worker and employer but the incidence of intractable disputes can be reduced if good procedures exist, and where the substantive agreements affecting wages and conditions of employment are coherent and rational.

The unofficial strikes in the mining industry arose from the wages system, as they do in most industries where the incidence of such strikes is above normal. Indeed, this was the evidence of the Ministry of Labour to the Donovan Commission and accepted by the Commission. It was certainly the case in the mining industry where, until 1967, half the unofficial strikes in this country occurred and were caused by the wage system of piecework. They usually arose from a claim for extra payment by the pieceworker to compensate for some anomaly or impediment to performance and earnings. It is not possible in any operation, certainly not in mining, to provide in the contract for all and every obstruction to piecework earnings. This is quite apart from the fact that I am strongly of the view that piecework, or payment by result wage systems, are completely outmoded in this machine age.

The preservation of the right to strike is vital to trade unionism but my experience, especially in the South Wales Coalfield, was that the unofficial actions thrown up by this wages system was not a condition to be preserved, unless one accepts that the tail should wag the dog, and that internal contention between sections of work-people affected but not participating in the action, strengthens militancy and unity!

The replacement of piecework in mining by a system of

measured time work and providing time rates based on the previous average piecework earnings for the district, has reduced the incidence of unofficial strikes to around a fifth of what they used to be. Industrial unionism makes possible the introduction of rational wage systems within which the relativity of rates and earnings as between man and man, and job and job can be determined by job evaluation and where the earnings gap, characteristic of the piecework era, is capable of control. The question that these developments pose to the opponents of industrial unionism is a simple one. Could such changes have been brought about had a dozen or twenty unions been involved and had to be consulted in the operation? The answer does not involve any cerebral strain.

The background factors and the full nature of the changes that have been brought about are worth some further examination, and reveal as many variations in conditions and practices as can be found anywhere. It is important to describe them to show that if they can be overcome and replaced by arrangements based on principles of equity and uniformity in one industry they can also be overcome in others.

Until April 1955, the wage arrangements throughout the industry were on a pit or district basis. The principle of a national minimum rate on the pit surface and below ground had been introduced by national arbitration awards during the last war. However, the rate and earnings for the job for both timeworkers and pieceworkers differed as between pit and pit and coalfield and coalfield. A particular job performed, for instance, in a pit in Nottingham, could attract a rate far in excess of the rate applying in South Wales or Durham. The disparities in rates and earnings were legion, and wide.

But out of this morass a structure for day wage men was constructed. It was a gigantic job, with thousands of local job names being compressed into national job names and job descriptions defining the main content of the job. A structure of five grades on the surface and four underground was devised, into which, by a process of rough job evaluation, each job was slotted for the purpose of applying the grade wage rate. In some areas and for some jobs the rates paid were higher that the initial grade rate into which the job had been slotted, and provision was made for the higher rate to continue on a personal basis but to be overeaken by the process of future wage increases. The only allowance payments which could be added to these national rates were those determined by national agreement and only for clearly defined special duties.

The principle involved in this agreement was that a nationally uniform wage rate was paid for the job no matter in what pit or coalfield it was actually perfomed; a wages system based on the content of the job and not the location of its performance. Although there are obvious disparities at least of degree, in the actual content of what are held to be identical jobs, the rough equity of wage distribution realized is a vast improvement on the anarchy that previously prevailed.

With such a major transformation in the wages system involving well over half the total labour force, it was to be expected that there would be a period of teething troubles, and a special dispensation was agreed to meet this. The intention was to move from the realization of a wages structure for day wage men to a structure for pieceworkers. However, the industry suffered a severe crisis of over-production between 1958 and 1960 and this created preoccupations for both the Coal Board and the Union that delayed the development agreed upon to rationalize completely the whole wages structure of the industry.

The early examination before revising the structure for pieceworkers revealed a state of affairs of such variation in systems of mining, wage payments, customs and practices that the task was felt to be an impossible one. The original intention was to achieve a national piecework agreement which would remove the main disparities of the old system. It was realized that if piecework systems had to be rationalized into a national structure, they would have to be replaced by time rates and that such rates would have to be related to piecework earnings if the change was to be accepted.

The new approach agreed upon was a piecemeal introduction of national agreements to meet the new coalface techniques and this led to discussions for a National Power Loading Agreement. Some idea of the mining system involved has already been given and for the present purpose it is sufficient to make clear that with power loading as with some other mechanical mining systems the output from a coalface is mainly determined by the performance of the machines, and not as in the old systems to which piecework applied, by sheer physical effort. The effect of this form of face operation increased the interdependence of the different work groups in the total operation of the pit and narrowed the differentials as between one occupational group and another in the exercise of skill, responsibility and labour effort. It was right and just therefore that if this narrowing of differentials in work performance was taking place, the wages structure should be revised to reflect this. This was the basis of the unions' case to the membership.

In most coalfields power loading was covered by a coalfield agreement and most of them included some element of piecework, with the result, of course, that there were wide disparities in earnings between pits and even within pits. It was understood from the beginning of the discussions that a national agreement would not in the first stage realize a national uniform time rate, and that variable rates related to district average power-loader earnings would mean that some men coming under the new national agreement would earn less than they had previously been getting. This was a great difficulty, and one that could have been easily overcome had it been possible to fix the new time rate to the highest earnings either nationally or in a district. Unfortunately, as the money needed for this was not forthcoming, a compromise was accepted; the wage was related to the district average, but with the provision that as wage increases proceeded those districts in the lower brackets would get larger increases, so that by the end of 1971 a national uniform rate would apply everywhere. In the case of six coalfields the rates were set above their average powerloading earnings in an effort to narrow the earnings disparities. Because of the acute difficulties experienced by the industry as a result of national fuel policies, the rate of progress towards this national uniformity was not as speedy as the parties to the agreement intended.

It was obviously difficult to achieve acceptance of this change from a system of wage determination that had operated in the industry for the best part of a century and which gave to certain workers the power to force up their own earnings. Men who felt they were due to lose earning ability and individual bargaining power had to be won over. Many were not won over before the agreement was introduced, and although a majority of members voted to accept, there was some sporadic rearguard resistance in certain pits for a period. No fundamental change of this nature could be initiated without some convulsions, and in the main the benefits of the agreement itself did most to overcome opposition. The guarantee that, no matter the geological or other disruption, at the end of the week the wage packet was assured, was a more powerful advocate that any of us whose advocacy was limited to words.

There is still a small area of operations in mining covered by some form of piecework, in most cases rather dubious forms aimed to provide certain level of earnings unrelated to any piece-rate calculation. This is now the subject of discussion between the Coal Board and the Union and when agreement is reached will result in a position where all wages in the industry are determined and regulated by national agreement.

What does this kind of development accomplish and does it benefit the workers in the industry? This is the vital question. Its great virtue for me is that it unites work-people in relation to their wages position, and this in contrast to the experience in mining of sectional strife exemplified in one coalfield by 'tit for tat' strikes. It offers the chance to reach and maintain a fair differential in wage rates between the different occupations. It provides a wage-negotiating system where for all the workers in the industry advance together, not section by section. It makes possible the benefits of increased productivity being shared by all and not as a concession to a privileged grade. It eliminates the discontent and disaffection created by the wide disparity of earnings, and reduces very substantially the incidence of sectional strikes. What is more, it makes possible an equitable and rational wages system for the whole industry.

These are the advantages that exist in an important industry; not just as a 'belief' but as definite achievements. They have been achieved not because the industry has only one employer, although this has undoubtedly helped, but more because the industry has had only one united organization of the work force, one union responsible to that work force in discussion with the employer. This is some proof from life itself, that industrial unionism is a far superior means of representing the interests of workers before the employers than representation being divided among a number of unions.

There are naturally problems thrown up by any process of change, and moving from local, or district, to national agreement is no exception. The price of achieving equity in wage distribution on a national scale means an intensified problem of communications within the union and industry.

Interpretations, amendments and improvement of existing agreements and the negotiation of new ones are constantly proceeding, creating the need for the closest possible relationship of the formal structure of the union with the membership. This means a perpetual system of communication between the Centre, the Area and the Branch so that the fullest explanations are made of any changes taking place. The standards embodied in National Agreements have to be maintained to avoid any possibility of local additions to the national rates. Any relaxation could lead to locally determined variations in payments and a return to the system based on where you do it, and not what you do.

In the miners' union, to assist in maintaining this close relationship between the membership and the leadership, a national journal was started, and it is hoped that this will be a medium of communication where information and discussion can proceed involving all levels of the union.

In all areas regular delegate meetings take place on a bimonthly basis where reports of the activities being undertaken and the problems being faced are given, and where the branches in turn can make their observations and table their views and in general participate in the discussion and formulation of policy. This maintains a fairly effective system of communication with the active members, but leaves a large part of the membership who do not attend meetings outside this formal communication system. It is hoped that the journal and the discussion that takes place in the pit among the active members will at least convey the main information to this section of membership.

One of the problems of the trade union movement as a whole is communications, and the bigger and more widespread the union administration the more remote the decision making centre appears to the man at the point of production. In an industry where there is one dominant union, with its basic organization covering each unit of production, the remoteness can be more easily overcome. At least, with only one source from which explanations and interpretations can emanate, the risk of distortion is less than where there are numerous sources, as in the case of multi-union industry.

The little experience we have in the mining industry of joint negotiation with other unions exposes the inherent weaknesses of such a system. We have experienced a union rushing to reveal the terms of an agreement before it has been finally settled, apparently in an effort to get the credit for itself, and the converse too, of a union delaying to accept a negotiated agreement in an effort to demonstrate that it was more militant than another involved in the same negotiation! It is equally the case that the more sources there are interpreting a joint agreement the more variations of its contents there are likely to be.

I have no doubt that the main criticism most likely to be levelled against national agreements based on time rates, will be that it eliminates the incentive to higher earnings by greater effort and this affects productivity. This could be the case where labour effort is the main factor determining productivity, and in such conditions piecework might be appropriate. But the argument is unsound where machine performance is the main factor determining output. Planned maintenance of the machine, and efficient organization in support of its operations, are far more important as measures to maximize productivity in mechanized operations. More often than not piecework is retained by management in these conditions to provide a smokescreen to hide inefficiency. Time rates providing high earnings are the best means of ensuring efficient supervision and management and, as a result, maximum machine performance.

It is hoped that the evidence of achievement by near industrial unionism in the mining industry will be sufficient to convince those who admit the weaknesses of multi-unionism, that industrial unionism eliminates those weaknesses, concentrates trade union power and, if it is possible to adopt it, is a superior form of trade union organization.

IS INDUSTRIAL UNIONISM POSSIBLE IN BRITISH CONDITIONS?

A thirty year interval between two studies by the TUC of the question, both resulting in rejection of the concept, tends to weight the answer towards the negative. My own answer is that industrial unionism as theoretically understood, or modelled to the West German sixteen industrial union structure, is not possible in existing British conditions. But having said that, I believe that British conditions – and by this is meant the existing union structure – can and must be changed nearer to vertical structures and that this could reach a stage over a period of years where even a structure near to that of West Germany would be practicable. This is not something to be left to the natural drift of evolution; there is a grave responsibility upon union leadership to steer a course towards the end of multi-unionism.

Many influences are operating now that were not present

to the same degree in the earlier studies. The economic and political changes, the assessment of which is the main thesis of this study, is compelling a re-examination of trade union structure. The re-examination is not only within the unions but is a major preoccupation of the dominant political parties.

It could well be that the central political issue upon which the next general election will be fought will be that of trade union reform. The next Government could be committed to giving power to the Registrar for Trade Unions to determine and control negotiating rights as a means of reducing and restructuring the unions operating in particular industries.

I am firmly convinced that a progressive reorientation of our unions towards a vertical industrial structure is vitally necessary and equally firmly convinced that it must be by measures voluntarily decided upon and applied by the trade union movement itself. To attempt to impose such changes politically would unleash a reaction and opposition far more hostile than anything so far demonstrated by unions against government interference. Interference from any outside body is objectionable and one naturally hesitates to put down ideas as to how the change suggested could be initiated and developed. Further, without a great deal more information on the present distribution of members of the different unions in multi-union industries, and the location of such members – and only the unions themselves can supply such detailed information – it is impossible to do more than indicate what could be the basis of an approach.

There are now around 25 million insured workers in Britain with around 9 million in membership of unions affiliated to the TUC. There is obviously in this huge gap a large area of trade union recruitment which could serve to cushion any membership transfers.

The Ministry of Labour evidence to the Donovan Commission, based on the 1963 position, published a table revealing that of the total affiliated membership to the TUC around 7 million were in eighteen unions and these were spread over the main industries and services. Another table revealed the wide disparities between union membership and the numbers employed, industry by industry. While it is true that a proportion of this unorganized mass is in small units and difficult to recruit, there is still a big element of lapsed members and non-unionists in the larger undertakings and services. This substantial recruitment potential is very important as a source available to compensate unions for possible loss of members by restructuring. A sustained and concerted recruiting campaign supporting the concept of industrial unionism is therefore a necessary accompaniment to the process of moving union structure nearer to an industrial basis.

There are 157 unions affiliated to Congress, but as the statistics show, membership is concentrated into eighteen of the largest. The remainder, although many of them catering for specialized occupations, have small memberships, many of them below 10 thousand. Their effectiveness is largely related to the special interests of the occupation or trade, and on major policy issues affecting wages and hours of work, tend to follow in the caravan behind the big battalions. Amalgamation and mergers to embrace these smaller organizations, while necessary to encourage and speed up, must at the same time preserve and cater for their special interests and, even more important, not be indiscriminate but conditioned to a coherent industrial pattern.

I consider that the policy of amalgamation and mergers over recent years, and those now under consideration, with few exceptions have been less concerned with any organizational concept and more concerned with enhancing the receiving union's power in relation to other unions. As already stated, fewer but more powerful horizontal unions can intensify the problems created by multi-unionism.

It is advocated in some quarters that agreements and pacts on 'spheres of influence' between the big unions is the answer to the problem. 'Agree on the plant, or occupation, or geographical locations, for recruitment and leadership, and limit competition and discord' is the advice. In my book this is a very 'shoddy compromise' and one that, at best, could only offer a limited and uneasy peace. It is just applying a soothing ointment where surgery is required.

The surgery need not be painful, although any kind of change is bound to disturb the *status quo*. Trade union leaders do not see themselves as tsars over big empires and certainly the rank and file do not see them in that role. Their concern is efficient and effective service to their members and their families, and if this is the primary aim the end justifies the means.

What is required in addition to an industry orientated policy of amalgamation and mergers is a shake-out of members from the big unions and their transfer to the union nominated as appropriate to be the main union for each industry under the supervision of the TUC. For instance, four of the biggest unions operate in the electricity supply industry. The industry is nationalized, so there is only one employer. It ought not to be impossible for these four unions to decide which of them is the appropriate union for the industry and to transfer the members from the non-appropriate unions to it. If this is too drastic an operation, the change could be by stages, the appropriate union being given main negotiating rights to which the others, after consultation, conform – the other unions stopping recruitment in the industry, so that by the process of natural wastage their membership declines to a level where bulk transfer is less difficult. Accompanying either method would be a joint recruitment campaign and a similar operation in other selected industries with a different one of the four unions chosen as appropriate, as for instance in gas, passenger transport or the motor-car industry. While the result would reduce the number of industries covered by these unions, their membership need not be significantly reduced.

This kind of inter-union sorting and redistribution of membership is practicable and, applied widely and over a period of years, could completely transform the trade union scene in this country.

The opposition argument to such a proposal will probably be that the policy would be against the 'freedom of the individual'. Freedom of choice in this area of human activity could result in even greater divisions than already exist. Such freedom does not apply now in a large number of industries due to closed shops and inter-union pacts. 'Freedom', I read somewhere, 'is the recognition of necessity' and it is the 'necessity' of changing trade union structure that is paramount.

One of the important reforms being sought by the trade union movement and the Labour party is an extension of industrial democracy, but how effective this can be in conditions of multi-unionism is open to serious question. It is a type of organization that inhibits any coherent form of representation in every industrial institution.

THE EXTENSION OF INDUSTRIAL DEMOCRACY

This subject has already been touched upon and is concerned with the extension of workers' participation in the control of industry. This was part of the perspective outlined in 1944 and was subsequently incorporated as a provision in the Nationalization Acts of the post-war Labour Government. Generally speaking, the form of participation in nationalized industries has been through formal and informal consultation between organizations in the industry and the National Boards. Also the odd trade union leader, usually near retirement, has been appointed to the Boards.

In 1968, at both the Trade Union Congress and the Labour Party Conference the issue was again highlighted. In Congress the debate centred on a resolution moved on behalf of the Transport and General Workers Union. The terms of the resolution call for some comment. The first part reiterates the claim made in 1944 to the effect that the extension of democracy in industry should enable workers to participate in decision making on matters affecting their working lives. This, it claims, would contribute to higher living standards, employment security and industrial efficiency. The second part calls for legislation to enforce this.

'It therefore welcomes the introduction of worker participation in management, and calls upon the Government to introduce legislation providing for Trade Union representation on the management Boards of all nationalized undertakings and other public authorities, and to expedite the study being made of this principle in the Transport industry.'

The Labour Conference adopted an executive document much on the same lines, the main supporting argument being that workers should have the same right and opportunity to influence policy in industrial affairs, in the running of industry, as is the case in relation to the affairs of local and national government.

The TUC submitted its views and proposals for this reform in its evidence to the Donovan Commission, but the Report of the Commission offers little encouragement to such a development, considering its own proposals for collective bargaining, being based on plant or company, adequate to meet any need in this direction.

Fundamental to this question is the acknowledgement that it must now be part of the purpose of the trade union movement to extend and guide the participation of work-people in the management of industry, and that this can be best accomplished by their representatives having some involvement in policy making and control. The principle of increased participation by workers was supported by the Labour Government and by some employers. In the journal *Management Today* of November 1968, there was this enlightened comment:

'Today's call for worker directors, and talk of industrial democracy reflects nothing revolutionary, but rather disatisfaction with a state of affairs which would not be endured in the larger political sphere. The last, the one we have now, is left over from the political credos, linked to the exclusive claims of property of a century ago.'

The right of workers to a voice in running the place where they spend a vital part of their existence is generally conceded, but the aspect of the policy likely to produce controversy will concern the form such workers involvement should take.

Both the TUC and the Labour Party proposals appear to be for an extension of the collective bargaining procedures to include discussion and agreement upon matters of management policy. The speech of the mover of the resolution to Congress put the concept in the following terms:

'One weakness in labour relations has been the traditional distinction between consultation – telling people – as against negotiation. The drawing of a line between what unions can talk about and what they can negotiate about is quite inappropriate when productivity bargaining has largely eliminated this decision. There is need to bring together the concepts of negotiation and consultation into a single channel of control so that we can no longer have a situation where one set of people consult about the smaller issues and another set negotiate on a different basis.'

In my experience in the mining industry consultation is neither 'telling' people nor is it a medium for negotiation.

What needs to be defined is what is meant by workers' participation or the extension of democracy in industry. There are obvious limitations which have to be faced in the interests of efficiency.

It has to be accepted that the final authority on what has to be done in the day to day running of any enterprise must be that of the person appointed to manage. Efficient management is a highly skilled job and while it is proper that the views of unions should be constantly sought on matters of general policy and taken into consideration, final authority and responsibility must be upon the manager. An industry cannot be run on the basis of a popular vote, or by majority rule. Neither is it practicable, in the event of disagreement on what action should be taken in a given situation, to refer the dispute to arbitration. In my view it is completely impractical to try to determine management policy in an undertaking by agreement between the parties, as is the case for resolving claims arising from conditions of employment. The most advanced stage of union involvement in management is in nationalized industries, and it would seem that the first step to be taken in the private sector would be to adopt the same system of formal consultative machinery.

In nationalized industries collective bargaining machinery is distinct and apart from the machinery of consultation. It is the union representatives who are involved in both procedures, so there is no duplication. To concentrate into one procedure questions affecting wages and conditions of employment and questions of management policy surely risks fouling up both. If the intention is to use management proposals to increase efficiency, or for some technical innovation as a bargaining factor for wages, by holding up the introduction of proposed changes, then the single channel procedure will have some merit. But usually it is accepted as reasonable for a measure intended to increase efficiency to be applied so that its effect on wages can be evaluated Without experience of the effect, the wage aspect would be certainly difficult to settle.

In my experience, where technical or other changes have been introduced and are held up to become the subject of barter, an antipathy to the changes themselves is invariably created which can hang on long after the issue of payment has been resolved.

A two channel system is well established in nationalized industry and it has worked reasonably satisfactorily. This is not an argument against experimenting with new forms of participation but it does seem sensible that this should be the system for general application in the private sector at least until some better system has been experimented with and proved.

But whatever form is applied, a host of problems will have to be faced and solved. The objective is clear, but its realization, when considered in relation to the union situation in most industries, is anything but clear. The proposal to incorporate consultation into a single channel of negotiations through the collective bargaining procedure is presumably dictated by a desire to accommodate the multi-union position. Some arrangement has been built up between unions to facilitate wage bargaining, and it is seen as less disturbing to build consultation into the existing arrangement. Such problems would not arise, of course, if vertical union structures operated. As it is, there is considerable doubt as to whether effective and worthwhile consultative machinery will work under conditions of multi-unionism.

At plant and company level there is but one management authority, and to be effective, the fewer the union organizations in machinery where there are discussions on planning and management policy, the better. The larger and more multi-union on the workers side, the less unified and coherent is their position in relation to management likely to be. The tendency, where this organizational spread exists, is for a division on 'who speaks for whom', and a variety of approaches to single problems based on the narrow vested interest of the crafts or occupations particular unions represent.

Although experience in the mining industry of representation through more than one union is limited, there was evidence of this sectional approach. For example, in discussion within the Consultative Committees on questions involving deficient organization of supplies, or poor maintenance of plant and machinery or matters critical of management, representatives of the lower grades in management structure incline to the defence of management. Where the discussion is on failures due to attitudes or actions of the workers on the production line or ancilliary services, representatives of the union catering for this section inclined to their defence, the result being a division and sometimes bitter conflict between representatives on the workers side. This was not serious in the union situation in the mining industry, but where representation is from a relatively large number of unions, this subjective and sectional approach to problems is certain to be intensified.

For this reason the incorporation of consultation on questtions affecting management control within the collective bargaining machinery is to carry into this new field of activity the same divisions and rivalries that frequently bedevil wage bargaining, as well as facilitating the extension of the area of 'barter' to questions of organization and planning within the undertaking. In industries and enterprises where this development of workers' participation is to be introduced for the first time, it should follow the pattern operating in most nationalized industries with a limitation on the number of unions to be represented. In fact until multi-unionism has been modified or eliminated, representation could usefully be limited to the union with the most members in the undertaking.

The consultative machinery in nationalized industries has been functioning for more than twenty years now, and very few changes in the system have taken place. After an experience over such a period there is obviously a case for experimentation to extend and deepen workers involvement in running industry. The obvious industries where new ideas could be tried out would be those where the existing forms of consultation have been reasonably efficient. The conditions of different industries will throw up different ideas, and it would be useful if an independent body, perhaps a Faculty at a University, made a study of existing methods and probed the minds of those with long experience in the processes of joint consultation, from which perhaps new and more effective forms would emerge.

Some extension could be usefully tried along the following lines. At plant level in addition to the formal and informal consultations now operating, a useful extension could be the involvement of leading union officers in the meetings between the manager, the planners, and other key management personnel, where plans for development are being discussed, or where staff are being briefed on plans already decided upon. By this means the special knowledge of union officers on the disposition of the work-people, and the labour problems that could be encountered could be assimilated at the early planning stages. While it is essential that the manager of an undertaking must have the final authority for decision making, the opportunity to influence decisions needs to be constant. Representatives of a union in a plant should be recognized as being as important a source of advice to management as those who are employed by them to provide that specialist service. In the larger undertakings longer term planning is decided at regional or national levels of management. The suggestion for involving union representatives in the initial planning consultations for plant level is equally important, and often more important, at the higher management level. Full time union officers could be brought in not just to be told of plans decided upon, but for their advice to be obtained and considered before plans are finalized.

The important principle being extended by this suggested development is that of 'accountability', both by management and union officers. Management is able to account for its decisions through the participation of union officers, and they in turn have to account to their members on the consultations they have been involved in.

The other existing method of involving workers and unions in decision making in nationalized industries is by the inclusion of union men on the Boards, either as full time or part time members. The weakness of this form of involvement is that the person so appointed ceases to have any organized connection with the trade unions and is not accountable to them. While such a person may, within the limits of the possibilities available to him within a Board, put a union point of view, this is not known to the union or workers in the industry. Indeed, too often the situation arises where the union man appointed forgets his past connections and becomes the advocate of policies he previously opposed. In its present form this form of participation is of little real value to unions or workers.

There is, too, a deep-rooted suspicion that men who accept such appointments have 'sold out' to the employers. Although unions insist upon such appointments to national Boards being made, the person who accepts appointment is popularly regarded as a 'renegade' or 'poacher turned gamekeeper', sometimes even by those who have been most vociferous in the demand for such appointments.

If the demand for an extension of industrial democracy is to be something more than a slogan for agitation, some union men must accept such appointments, and as an old colleague of mine used to say, "if we want this, then we should put our best men there'. The weakness of the present system is the isolation and exclusion of the men appointed from the trade union movement. Participation in management control that is accompanied by non-participation in the trade union movement is a step backwards, not a step forward.

Any extension of appointments to national Boards or experiments for change must include the principle of accountability by the union man appointed to the unions and workers in the industry concerned. It is largely the absence of this in the present system of appointments that gives rise to condemnation. The ideal position would be the appointment of union men for a specified period, subject to recall by the union to which they were previously attached and responsible for keeping the union informed on the Boards' main activities and representing within the Board the interests of the union and workers. But there are practical difficulties. The members of national Boards are invariably given responsibility for particular fields of activity suited to their special qualifications. This would not be easily operated in the case of workers' representatives who may have no special technical qualifications and whose purpose on the Boards is a general one rather than to perform a specialized function. Board members are usually appointed for five year periods and this could equally apply to union men, who would need to be full time officials who could return to their former posts at the end of their period of appointment, if there was no agreement for this to be extended. It would seem that such appointments should be additional to the functional composition of the Boards, with the person appointed being responsible for liason with the unions operating in the industry. There would be problems related to salary and conditions of appointment, but experiments are proposed, possibly in a selected nationalized industry, to try out a new idea for a period to assess whether the idea is workable and advantageous.

An alternative experiment could be on the basis of part time appointments of both union officers and workers to the national Boards as well as the regional Boards. The appointment would be limited to attending meetings of the Boards when policy matters are being discussed and would mean, at most, only one meeting a week. The part time method is readily practicable and would provide a system of communication between the leadership of the national union and the Board of a more intimate nature than that prevailing under the existing formal system. It is equally practicable at regional level for men from the production process to participate in regional policy discussions which would facilitate the interchange of ideas on the problems of the industry between workers and management.

Whatever the experiments with the form of extension of workers participation, and the foregoing suggestions are only ideas for experiment and discussion, the process will mainly be one of feeling the way forward by trial and error.

The end product will be an improvement in the status of workers in their place of employment, and an extension of their ability to influence operations. It means an extension of management accountability to their employees for their stewardship as managers and employers, and at least a step forward in democratization of industrial management.

It does not mean workers' control or anything near it, any more than Parliamentary democracy means workers' or peoples' control over political affairs. A great deal of spurious nonsense is written about 'democracy' and 'freedom'. They are not absolute conditions and never can be and their definition and application in different countries, under different conditions and by different people, can be poles apart. The absolute in both concepts is for the individual to do as he pleases, free from any rules or laws or consideration for others, usually described as a state of anarchy. The ideal concept of freedom and democracy is to devise a set of rules and laws which dovetail the interests of the individual or section with the interests of society as a whole, and the extension of democracy in industry is one step in that direction.

The term 'workers' control' is used equally loosely and appears to me as a throw-back to the earlier and less mature days of the socialist movement in Britain when 'syndicalism' appeared to mean the same thing.

It is an unworkable proposition under capitalism, and in

a socialist society the term 'workers' would have to be given the widest possible definition, as would the term 'control'. It seems to me that such control is impractical as a piecemeal development within capitalism, and is attainable only by political power over all economic and social activities.

THE TRADE UNION CONGRESS. DOES IT LEAD?

If, as some trade union leaders claim, the Trade Union Congress is the parliament of the organized workers, it would logically follow that the General Council is its cabinet. The parallel, however, is not a very real one. The authority vested in Parliament and its institutions of Government, including the Cabinet, is in no way a model for these two organizations of the trade union movement.

The decisions reached at annual congresses after useful debate on the many resolutions that form the agenda, carry very little authority and commitment over the organizations represented. Recommendations from the General Council on important policy matters may be formally approved by Congress but may not be regarded as binding upon the affiliated unions. Indeed, the resolutions and decisions reached, particularly the terms in which they are presented, are not accepted and applied by the General Council as instructions to be faithfully and strictly carried out. Congress has been told on more than one occasion that the General Council must always be free to apply their own judgement and interpretation of the terms and policy content of resolutions.

Having participated in Congresses over a period of thirty years, latterly as a national officer of the miners' union, I must confess that the decisions of Congress had little significance for the union, despite the lengths some union leaders who were members of the General Council would go to get their unions committed to the policy the council was pushing. The promoters and supporters of resolutions would know from experience that a favourable vote would carry little weight with the unions who may be in opposition, and not necessarily any considerable weight with some who voted with them. Congress decisions were not really decisive in shaping the policies of individual unions. A decision reached by a majority vote has therefore been little more than a record of the attitude and outlook on a particular question at a particular time.

The positive aspect of a debate and vote is that the issues involved will have been clarified for the unions and public. Knowledge and understanding will have been advanced and the approach of the various unions to particular questions and their reasons made known.

With all its limitations Congress meetings are important as a forum where unions can discuss common problems, even if the action in relation to them is sometimes sectional. It helps to build and maintain trade union solidarity.

Although this has been the general background in more recent years, a growing recognition has been noticeable, a recognition that Congress and the General Council has filled a role in national affairs that could not be undertaken by unions separately; and this has led to a position where decisions taken collectively, on certain questions at least, carry a sense of commitment by the individual unions. Nevertheless, the TUC still lacks in power and authority.

This reluctance to transfer power is no doubt conditioned by the wide variations in outlook and interests embodied in the movement. All the political postures to be found in the labour movement generally, between the extremes of 'left' and 'right' find expression in the unions, in some places with the added ingredients for good measure of a sprinkling of liberalism and toryism. In addition to these political outlooks, there are differences of interest, thrown up by association with particular industries which often put unions in positions of conflict with each other on questions of industrial policy. All these differences colour the debates and activities of the movement which may in times of crisis sharpen into open conflict. Trade unions are mass organizations of work-people and must bring together workers irrespective of political or religious outlook. Tolerance of these differences is paramount, with the result that unions operate not as political organizations but as organizations pursuing aims which represent common cause of all their members arising from their conditions of employment.

Against a background of this rather loose relationship of affiliated unions to Congress, the remoteness of Congress and the General Council from the rank and file trade-unionist is understandably not inconsiderable. The trade union movement of Britain has grown out of the close local organizations of a century ago to the district and national organizations of the present. From my experience in the miners' union the trade union centres regarded by the ordinary member as his union headquarters, rarely included London, and were mostly the unions regional headquarters.

If this can be the attitude to a national union, it is reasonable to assume that the TUC as such is regarded as even more distant and probably as a more dubious trade union asset. If a centre of a national union could be regarded as a rather doubtful asset or as a body to be blamed when difficulties arise, it is unlikely that the TUC, even more removed from direct contact with members, will be regarded differently. In present circumstances both conditions are disabling factors badly in need of remedying.

The fact is that Congress really should now be the parliament of the organized workers with the General Council acting as its cabinet in leading and guiding its activities. In the councils of the nation the Trade Union Congress has achieved a status which needs to be backed up with the authority and power its status warrants. It has to speak for the trade unions and its statements must be known to those to whom they are addressed as carrying the full weight of a united Congress. This is imperative if it is to be listened to and heeded. Power with the employers and the State is centralized and it would be plain daft for the trade unions to carry on as if these changes had not occurred. Power in the trade union movement needs to be strengthened and concentrated more in the centre if a balance of power position more favourable to the members of trade unions is to be realized.

The present state of affairs must obviously be changed if the TUC is really to act in relation to the employers and

Government as a powerful and authoritative force, whose pronouncements are to be regarded as decisive, and whose representations are to be effective. To achieve this power and authority at the centre of trade union activity will require affiliated unions to sacrifice some of their own autonomy. There must have been many occasions when the effectiveness of General Council members in the important councils and committees was weakened by the knowledge in the possession of employers' and Government representatives that they could not commit all the unions to the policies they were pressing. How many times, one wonders, might there have been joint agreement or understanding with other parties to discussions had they been assured that the TUC representatives were really speaking for the trade unions collectively? Their power to reach conclusions and decisions on matters they pursue is inadequate for the tasks they have to undertake.

The suggestion inherent in this criticism is not that of giving arbitrary powers to the General Council, or of creating a bureaucracy. It is a question of policy decisions based on a majority vote at Congress carrying a greater measure of commitment for the affiliated unions. The trade union movement in Britain is to a large extent a prisoner of its own history and traditions, but it is a prison from which a 'break-out' is not only possible but completely justified. In a number of countries where history and tradition are radically different, the central organization of trade unions has the power to act for the movement as a whole and is able to exercise and command discipline from affiliated unions. In these countries the problem of vesting power at the centre has been resolved, and matters of national policy involving discussion with the Government or central organization of employers are left to the trade union centre with power to take decisions when necessary.

In this country we are rightly proud of our traditions, but when they are allowed to operate as barriers to change to help solve the problems of the present they cease to be the source of experience and wisdom. However, there is some evidence of a slow move towards facing realities and increasing the power of the Trade Union Congress. The Special Congress of June 5, 1969 was historic not only as a rare event, but by approving a recommendation from the General Council to give it greater powers to intervene in inter-union disputes and unofficial strikes. This represents a beginning of a development that must be continued.

The two issues to which this new power is to be directed – and restricted – are relatively unimportant when set against the necessity of increasing TUC power to match that now represented by employers and Government. The issue of unofficial strikes arises in the main from wages structures that are no longer relevant to this machine age, their incidence being greatest in those industries where payment by results operates. Revision of these structures would seem a more effective remedy.

The inter-union disputes are a by-product of multi-unionism and will not be solved by the measures proposed, although their worst effects may be avoided. In fact, the more powerful unions are more often than not involved in these disputes, and sanctions or threats of their use could result in more damage to the TUC than to big unions! The TUC will have to rely on its powers of persuasion and the degree of loyalty unions have for it, in the future as in the past, to effect settlements.

The importance of the decision, however, is not whether the TUC can put sanctions into effect, but that affiliated unions were almost unanimously in favour of giving this power to it. Earlier, the traditional attitude of unions to Congress decisions has been described as that of a declaration of policy carrying little sense of obligation upon them for its implementation. But strangely enough, on the major questions that have been spotlighted for intense public discussion as well as heated internal debate, a strong sense of obligation not to 'rock the boat' is evoked even by the unions who declare their opposition. There seems to be a greater measure of loyalty to majority decisions on issues of major policy than is the case in relation to matters which, although important, are not the subject of such keen public interest. On matters which are 'hot politics' as in the case of Incomes Policy which produced the TUC's own wage-vetting policy and more recently on unofficial strikes, unions who were opposed to the policies accepted by the majority, as well as those whose support was half hearted, fell into line with the adopted policy. This is certainly the case in respect of the wage-vetting policy where all unions conform to the procedure laid down, and will undoubtedly be the case towards all policies adopted which emerge from the pressures of national politics. The concern of the individual unions is to avoid internal conflict in such circumstances, to maintain unity in face of strong political pressures, and to subordinate their independent viewpoint to that of the collective opinion.

The logic of this situation is that when the trade union movement as a whole is facing a concerted attack, or what it considers to be an attack, it will close its ranks, despite differences that exist within it, to meet the common danger. This is a very strong positive quality, and one dictated by the urge for self preservation, and the recognition that this is only possible by closing ranks and centralizing power, if only in respect of the particular activity under attack.

In the two examples referred to, the response was prompted by a clear appreciation of the situation which necessitated it. The confrontation by the Government both in respect of Incomes Policy and their proposals for dealing with unofficial strikes, left no room for doubt about what the unions were facing, required no deep analysis, and was for them little more than a question of deciding whether they were 'for' or 'against'. With this kind of background, obtaining agreement to give power to the TUC to act more decisively in the matters under attack was possible, although it was reluctantly conceded by some unions and after much heart searching.

But is this same response possible to a more general, less dramatic but far more serious and challenging confrontation? Is it possible to get the same clear appreciation of the need to close ranks and increase the power and authority of the trade union centre in Britain, to match the power of the institutions which are decisively shaping the conditions within which the trade unions function? These are vitally important questions, the answers to which can effect for good or ill the future of trade unionism in this country. In a brief five minute intervention on the General Council Report to the 1968 Congress dealing with union structure and mergers, I tried to outline the two changes I considered important for the trade unions to undertake to meet the changes in environment, and which have been more fully described in this study. The two changes required I am more than ever convinced are, one, to move as quickly as possible to an industry based union structure and, two, to increase the power of the TUC and General Council. They are two parts of a single process aimed at making trade unionism strong and effective in the conditions of modern Britain.

4

The Power and Constitution of the TUC

AN examination of the objects and constitution of congress and the duties of the General Council exposes some of the limitations of its authority. The objects are naturally broad and general and directed to 'generally improve the economic or social conditions of workers in all parts of the world, to assist other organizations with similar objects, and to assist in the organization of all workers eligible for membership of its affiliated organizations'.

This last reference to assisting in the recruitment of those 'eligible for membership of affiliated organizations' is certainly an object very difficult to pursue in conditions of multi-unionism, where eligibility can be to any one of a dozen or more unions. It means that in the prevailing conditions the TUC is rendered impotent in membership recruitment by this impossibility of defining eligibility in terms of organization. To try to do so would in most instances involve the risk of being accused of taking sides as between one contending union and another. If the pattern of union structure was industrial this limitation would be removed and the way would be opened up for a sustained all union campaign to build up membership.

To a lesser extent there is a similar limitation in pursuing the object of improving the economic and social conditions of workers. Within the country there is the constant consideration of not cutting across policies being pressed by individual unions, and being restricted to assisting and not initiating policies except of the most general kind. In fact, the TUC is probably less inhibited in pursuing this aim internationally. To really fulfil these objects in modern conditions, policy decisions of Congress, based on majority rule, must supersede the policy decisions of individual unons and must be accepted as doing so by them. This, I consider, is the first change that unions must accept, and if this requires a change in their own rules and constitution, the change should be made. The TUC must no longer be regarded as a loose federation; individual unions must recognize that their own power is now puny without it.

It is precisely in relation to 'economic and social conditions' that increased power to the TUC is necessary. It is not just a question of making and presenting an analysis of current trends and difficulties, but of recommending policy and being responsible for its operation, once accepted. The difficulty with individual unions arises not so much from the statement of general policy and intent, as when this policy is given specific meaning to an industry, or a set of wage negotiations, or social and industrial reforms that may not be radical enough for particular unions. It is because these issues are now questions of national politics, involving Government policy, that the role of the TUC is decisive and that of individual unions is one of assisting and supporting. In these matters roles have been reversed.

It is remarkable that these considerations do not arise in a whole range of TUC activities. Policies in respect of education, social security, industrial injury and disease, and similar questions are accepted by the unions generally as questions for action by the General Council and Congress. The reason must obviously be that unions recognize that reform in this field involve Government policy and it is therefore proper for the organization representative of all the unions to take responsibility for prosecuting the policy aims decided upon by Congress. It is strange that against this long history of General Council monopoly in dealing with these matters that some union leaders can still brand as 'outside interference' any participation by the same body into industrial affairs. Union sensitivity is limited to intervention by the TUC into industry, into wages and conditions of employment, regarding this area as their own exclusive preserve. No one in their right senses would propose that the TUC should take over responsibility for negotiating specific wage claims, or reforms special to a particular industry. This obviously is the job of the individual unions. But the TUC must be given some power to prevent policies being pursued by one union that can embarrass or weaken the bargaining position of another in the same industry.

A wages policy, as stated previously, is not only a matter of securing a certain level of increase, but of making possible some orderly progress so that one section is not advancing at the expense of another and without regard to the interests of working people generally. What is required is an acceptance by unions that matters arising from industrial relations concerning, or subject to, decisions of the central organization of employers or the Government, shall be matters for action by the General Council in exactly the same way and with the same authority and power as by tradition has applied to matters like education and social security. This is the only effective way by which the objects of Congress and the trade unions can be pursued now and in the future.

The duties of the General Council as set out in the Constitution relevant to this aspect of activity are to be found in Rule 8. (d).

'They shall promote action by the Trades Union movement on general questions, such as wages and hours of labour, and any matter of general concern that may arise between trade unions and trade unions, or between the Trade union movement and the Government, and shall have power to assist any union which is attacked on any vital question of trade union principle.'

It has to be acknowledged that the General Council has not been particularly lively in 'promoting action', regarding itself more as a co-ordinator of union activities on the matters referred to in the rule. Economic reviews and general discussions with the Government on the questions thrown up by the review, cannot be regarded as promotion or initiation of 'action' no matter how moderate the interpretation of action might be. Undoubtedly, this attitude of negative leadership has contributed to a certain lack of confidence by unions in the ability of the General Council to deal with questions of industrial policy which the proposal to increase their power and function in this field would entail.

I recall a situation early in 1962 when the miners' union met the officers of the TUC to consider some action against Government policy which was frustrating a wages settlement in the mining industry and at the same time in a number of other industries.

A general offer had been made by the National Coal Board, but the union was pressing for a further 3d a shift on the rates of the lower paid men. There was good reason to believe that the Government had directed that this very slight improvement should be refused. The three national officials of the union met with the chairman and secretary of Congress on February 12, 1962, and pressed for a conference of other unions facing a similar situation to decide upon some form of common action. The report of the General Council to the Congress gives the reasons submitted to justify the Conference in the following terms. (p. 247)

'On the subject of what should be put to the conference, the NUM representatives stated that one-day token stoppages had been arranged in engineering, and that rather than unions in other industries acting similarly there might be co-ordination of such activities for a single twenty-four or forty-eight hours token strike.'

The reply of the General Council, following consideration of the Unions proposal by the Finance and General Purposes Committee:

'It was therefore decided to inform the NUM that, in the opinion of the General Council, the proposals put forward by the Union for action of a general character would not be effective and could not be usefully pursued. If however, the mineworkers wished to specify to the General Council any particular action which they wished to be taken relative to the wages negotiations in the mining industry, the General

Council would render such assistance and use such influence as they could.'

In this way neither promotion nor co-ordination of action was considered worth pursuing. The offer of help in the negotiations, while being well meant, was not taken very seriously by the union who took the view that it was unlikely that anyone from the TUC would be better equipped for the particular job than their own negotiators.

This reluctance to co-ordinate activities is not confined to the General Council. Early in the campaign against pit closures the miners' union arranged a meeting with the three railway unions who were facing a similar process of rapid contraction. The intention was to get agreement for a joint campaign. The meeting took place but most of the time was spent listening to rather bitter recriminations by the representative of the ASLE & F against the NUR related to wage negotiations in the industry, But even despite this there was no disposition by any of the rail unions to join forces against these Government sponsored policies.

One explanation for this attitude of the General Council is undoubtedly the prevailing character of union structure and its reflection in the composition of the General Council. The union leaders who make up the Council are not only doubtful about increasing its power because of the effect it might have on their own positions, but because of their competitive relationship in industry they tend to distrust each other. The distrust would not be lessened by engaging in a joint campaign on a common demand. One union would probably outshine another, and would probably set itself out to do just this, and union personalities would stand out in contrast one with another to advantage or disadvantage. This same consideration exists when attempts are made to initiate joint activities by a group of unions facing a common problem. These attitudes are of course, accentuated by political prejudices. The problem therefore is not just one of 'constitution', but involves strong subjective attitudes thrown up by trade union structure and the relationships it produces.

POWER TO SUSPEND

The power to suspend a union for 'conduct detrimental to the interests of the trade union movement' is written into the Constitution.

'If at any time there appears to the General Council to be justification for an investigation into the conduct of any affiliated organization on the ground that the activities of such organization are detrimental to the interests of the trade union movement or contrary to the declared principles and policy of Congress the General Council shall summon any such organization to appear before them or their appropriate Committee by duly appointed representatives of such organization in order that such activities may be investigated. In the event of the organization failing to attend, the investigation shall proceed in its absence.'

The rule further gives the General Council the authority to direct any offending organization to discontinue such activities forthwith, and if they disobey, to suspend them immediately. In such an event the next following Congress would have the right to ratify or reject the report.

The power represented in this rule is very comprehensive. It covers detrimental conduct which could reasonably be interpreted as malpractices in the conduct of union affairs; acting contrary to declared principles, is a little more difficult to define, but an extreme example could be one union organizing 'blacklegs' to break an official strike by another; and finally, 'contrary to declared principles *and* policy' makes the disciplinary power of the General Council very wide, if principle and policy are taken to have roughly identical meaning for the trade union movement. The position would be more decisive if the 'and' was substituted by 'or' if it was intended that acting contrary to 'policy' could invoke the machinery for discipline. I have no doubt that a lawyer could make a verbal feast around the difference in the significance of these two small words, but it is to be hoped that for most trade unionists principles and policy are synonymous.

If this is the case and the intention of the rule if strictly applied, then it could follow that an affiliated union which acted contrary to a policy decision of Congress and refused to discontinue such activities could be suspended. Set against the background of tradition the implementation of such an interpretation would be suicidal. Imagine a situation where Congress has decided that wages increases for a given period should not exceed 5 per cent and this was accepted and applied by a majority of unions except one which refuses to call off a strike for 7½ per cent and this union was suspended. The support among trade unionists would rally to the suspended union who would seek to force a lifting of the suspension. Disciplinary action of this kind, no matter how strongly actions contrary to Congress policy may be condemned, is impracticable, except in circumstances where serious malpractices have been proved.

Acceptance and implementation of Congress policy must be based on consent and the operation by affiliated unions of a self-discipline. Only by persuasion can the General Council and brother unions act to influence the policy and actions of a union refusing to carry out agreed policy.

I have had some experience of implementing the sanction of suspension against a small association within the miners' union. It solved nothing but instead created new and difficult problems, in that the section involved became a permanent irritant to the normal operating of the union and the industry in the area concerned. In the end discussions had to be initiated with the suspended section and terms for their reinstatement agreed upon which were more favourable to them than those made before their suspension.

The question is not that of strengthening the constitution of Congress or the rules governing the authority of the General Council. They have an authority now they cannot operate because the consequences could be worse than the condition the discipline is expected to remedy. There is no merit in possessing powers that cannot be used. What needs to be strengthened is the willingness of affiliated unions to accept and operate Congress policy and, if constitutional changes are warranted at all, they could be in individual union constitutions to provide that where questions arise which involve national politics, the policy decisions of Congress will supersede those of individual unions, if and when differences exist. The essence of the case being argued at this point is that while there are serious weaknesses with the General Council and Congress, they are not weaknesses of constitution and rule. All the arguments for strengthening the constitution used by politicians and publicists mean only one thing, sanctions. The trade union movement is a voluntary association of members and organizations within which sanctions can only have a detrimental effect.

THE GENERAL COUNCIL AS A REPRESENTATIVE LEADERSHIP

The basis upon which election to the General Council is organized is conditioned by the anarchic state of union organization generally. Until a coherent industrial trade union structure is realized, the basis of election will be unsatisfactory and inadequate. The 1968 Report to Congress sets out details of the present position. Industries, trades occupations are classified into nineteen groups. The number of unions in each group and the membership covered shows wide disparities. In Group 1, Mining and Quarrying (although now no quarrying membership is included); there are three unions with a combined membership of a little over 400,000. Representation from this group is three, but is to be reduced to two by the decision of this Congress. In Group 3, Transport, other than Railways, there are nine unions with a combined membership of over a million and a half. Of this membership one union includes over a million and a quarter, while another union covers only 261. Representation from this group was increased from three to four, and are from the big unions only. The reorganization represented by this change in representation is superficial and does not touch the main sources of dissatisfaction.

The Report of the General Council explaining these limited proposals for change offers the following reasons as justification:

'The Group system is intended to produce a General Council which as far as possible mirrors and represents the Trade Union movement as a whole, and the need to preserve a fair balance of different interests and experience has always taken precedence over the simple approach of distributing seats on a purely arithmetical basis. On the other hand, account must be taken of changes in union membership and of trends of employment in industries and occupations covered by the Groups.'

The above extract from the Report to the 1968 Congress implies that the system of electing members 'mirrors' the prevailing union structure and that any basic change can only follow a change in structure. A new Group covering 'Technical, Engineering and Scientific' occupation has been set up, but since technical, engineering and scientific occupations are to be found in every industry, the object is obviously to give representation to a group of unions previously smothered in another group. Reorganization like the original organization is based on purely arbitrary standards, and determined more by considerations of expediency than of principle, and while some anomalies are eased others are perpetuated. Take the important industries covered in Group 14, 'Glass, Pottery, Chemicals, Food, Drink, Tobacco, Brushmaking and Distribution'. This Group covers fourteen separate unions and has two representatives on the General Council. The industries covered are widely dissimilar and no two representatives can cover all the interests involved.

While trade union structure remains as it is there can be no method other than an arbitrary one for election to the General Council, and because of the huge disparities in union membership within the same group, a farcial element of democracy in the voting. The 'no contest' or 'reserved' seats, as they were described by the late Bryn Roberts, are usually the result of a recognition by smaller unions that their nominee would stand no chance of being elected, unless he was sponsored by the 'big brothers'. It is also generally acknowledged that the present system of voting involving Congress as a whole, guarantees continuity of election by the favoured few, under the rule of the 'old pals act'. In fact a compact between two or three of the bigger unions could determine the composition of the Council, and the fate of current members.

An election system that is generally ridiculed and acknowledged to be a rather phoney application of democracy ought to be reviewed and changed if only to maintain an image of propriety and dignity in the conduct of union affairs. An election system that would be condemned if it applied in an individual union is not one to be praised when it applies to the important function of producing the best possible central leadership.

The present system permits unprincipled ganging up to keep from being elected union leaders who are looked upon with disfavour by the dominant group, although from the standpoint of ability and other considerations their election would be to the advantage of the movement.

The proposal has previously been made to change the present system of election to permit voting within the groups to decide representation, and not the vote of Congress. This proposal is only another makeshift arrangement and would still perpetuate many of the anomalies of the present system. Change in representation to the General Council requires change in union structure, and even then some grouping might be necessary, certainly in the early period of such a change. But the grouping could be made more coherent by including industries that have much in common and an affinity of general interest. It is, I realize, a rather monotonous theme, but whatever aspect of trade union activity is examined and the need for some change is exposed, the first prerequisite to change seems always to be the restructuring of organization to an industrial base. This position also arises from an examination of the General Council as a representative leadership.

WHAT IS TO BE DONE?

There appears to be no short cut to the kind of reorganization that is required. Until a start is made towards industrial unionism, changes affecting the General Council aimed at making it a more effective leader seem likely to be piecemeal and less than the situation really requires. In the intervention already referred to which I made on this whole question at the last Congress, the following suggestion was put forward:

'I would suggest the setting up of a small committee by the General Council, comprised of people who have no vested interest in the present structure, to prepare a blue-print for revision both of the structure and the power and functions of Congress and the General Council.'

The suggestion of setting up a small committee to prepare a blueprint as a guide to general reorganization is based on the belief that the job cannot be left to sub-committees of the General Council no matter how well intentioned they may be. It is difficult enough for detached people to be wholly objective, but largely impossible for those who still hold office in the trade union movement and are subject to the pulls and stresses associated with their position.

In the section dealing with trade union structure suggestions have been made as to how some change might be brought about. It is not suggested that other methods cannot be found and which might be more successful. What is more important is that a study of all possible methods is made and put forward for discussion.

In examining what steps can be taken to strengthen the General Council there must first of all be a general acknowledgement that it is necessary. The purpose of this study is to try to show that it is. If, as is strongly argued in these pages, the new relationship of forces warrants a substantial transfer of power to the trade union centre, the question arises as to whether the General Council and its existing staff are adequate to discharge the responsibilities such increased power will place upon them.

There has already been an increase in staff employed at Congress and there may be a case for some further increase, as the work of Congress is extended. But however competent the staff may be, they operate under the direction and guidance of a full time General and Assistant Secretary. The sub-committee could usefully determine whether further full time trade union officers should be elected to meet the new responsibilities and duties. The set-up that has operated down through the years has been largely that of a one man band and might have been adequate in the conditions of an earlier period. My own view is that such a set-up at the centre is as outmoded as trade union structure itself.

The strengthening of Congress leadership by the election of a number of full-time union officers to become an inner cabinet of the General Council would provide a core of experienced leaders with the ability to effectively meet the confrontations that the new situation will continuously present.

They would need to be elected from the most able of the national trade union leaders since part of their job would be to advise and guide individual unions about the problems delegated to the General Council. Obviously, since they would have to give up their jobs in the unions to take up office with the TUC, their appointment would need to be permanent and not subject to periodic election. Such elected officers would become members of the General Council with the right to speak and advise but without the right to vote. The purpose of this suggested restriction is to avoid creating a trade union 'junta' within the Council.

The arguments against this suggested development are along the lines of 'too many cooks spoil the broth' and the most capable union men would prefer to remain with their own unions. The answer to the first argument is obviously that with only one cook there will be no broth to spoil for that part of the demand that exceeds his capacity. On the second argument it has to be acknowledged that in the early stages of such a change some reluctance on the part of union leaders is to be anticipated. The salaries would obviously have to be attractive, but the main inducement would be the knowledge that the centre of power was no longer at the head of a particular union but at the head of the trade union movement as a whole.

The suggestion has been made that the position could be met by the election of a full time President. I have always had doubts about the wisdom of having two full time officers of equal status at the head of a union. The only thing equal in such a set-up is usually confined to status, and in respect of the things that matter it is the 'differences' that stand out. The General Secretary must carry the main responsibility, with a part time President, and full time elected officers and staff acting as his aides and general staff.

It would be useful, too, if this detached committee would examine the whole question of communications and publicity within the trade union movement. The production of a monthly bulletin with a very limited circulation and infrequent unorganized press conferences, is totally inadequate for an organization of the importance of the TUC.

If there is any organization which needs a full-time, top-class press officer, it is the Trade Union Congress. It is the most maligned and misrepresented organization in British society, yet its power of reply is that of a whisper compared to the roars of its critics. The old adage that 'virtue is its own reward' may sound fine as an abstract piece of philosophy, but in practice in the the kind of world in which the trade unions operate it can be self destructive.

There is an acute need for planned and well organized communications both between the trade union centre and the unions and members, and the general public. This is no longer the job for the layman, or the union leader, no matter how able he may be in that capacity. It is a job for the trained specialist and the time is long overdue for the General Council to take steps to remedy this serious deficiency.

There are six main proposals or suggestions for consideration arising from this brief study of the functions and constitution of Congress and the General Council.

First, because of the changes that have taken place in the general environment of trade union operations the role of Congress and the General Council becomes decisive. It is therefore imperative that the power and authority of both should be strengthened.

Second, that the policy decisions of Congress on questions of general policy affecting industrial relations should be accepted as binding by affiliated unions. This does not mean participation by the TUC in specific wage negotiation but is concerned with aspects of industrial problems in which the State and the central organization of employers are involved.

Third, the General Council to be strengthened by the election of a team leadership, made up of experienced union leaders, whose job would be to guide and mould the movement to meet the changes already effected and those to come.

Fourth, to examine the possibility of a more genuinely democratic system of election to the General Council.

Fifth, to appoint a full time top ranking press officer to organize the communication between the centre and the members and to the general public.

Sixth, to consider appointing a small committee of experienced people to examine these and other proposals with a view to presenting a blueprint to the General Council and Congress as a basis for restructuring the trade unions and increase power at the trade union centre.

5

Strategy for the Future

ANY attempt at reviewing strategy and tactics for the future must make some assessment of the perspectives before society and their effect upon the environment of trade union operations. Two of the main factors influencing the outlook have already been referred to in some detail, and need no further elaboration now. They are, the continuing movement to larger commercial and industrial organization in the private sector of the economy, and the continuing and probably increasing involvement of the State in economic and industrial affairs. These two factors will exert a growing and permanent influence on the social and economic environment of the future. The trends in this direction are already well established.

The first factor to be examined is that of change in the character of the labour force especially from the standpoint of its influence on the composition and outlook of the trade union movement. The 1965 National Plan – this Plan is being quoted not because its estimates are accepted as accurate, but because it indicates trends – forecasts that the national working population will in absolute terms continue to increase. In addition to this increase overall, there will continue the redistribution from the contracting industries to the industries and services that are expanding. Between 1965 and 1970 it is estimated that public administration, health and education will require an additional half million employees. Of the 26¼ million in employment in 1970 the growth factor in the redistribution of employment is to the non-manual occupations.

The sectors releasing labour are the manual industries in the main, such as mining, railways, textiles, agriculture and shipbuilding. In fact the run-down in the labour force in these industries is proceeding far more rapidly than was estimated in the Plan. Mining is an example: the Plan forecast a reduction to 480,000 by 1970, revised by the Ministry of Fuel and Power in 1967, down to under 300,000, but the total was in fact down to 328,000 by the end of 1968. On present trends it will be probably down in the region of 280,000 by the end of 1970. The Plan, while portraying the trend, greatly underates its rapidity in the contracting industries. It underestimates too, the displacement and redistribution rate flowing from mergers and takeovers and the process of concentration which these developments bring about. However, the significance of this redistribution of labour is both quantitative and qualitative. With the changes in the numbers being employed in the various industries and services there is also a change in the ratio of manual to non-manual occupations and this will obviously have an effect on the quality of trade union membership for the future.

The other feature of the growth of the working population is the increase in the numbers of women in employment. Before the war fewer than 5 million women were in employment, but it is now estimated that by 1972 there will be around 9 million. Thus the composition of the labour force will radically change to include substantially more women, more 'white collared', administrative, technical and clerical employees, alongside the severe contraction of the older and heavier manual industries.

The effect of this change upon the trade union movement is bound to be significant. The decline in employment is in the industries which stimulated and fashioned the growth of the trade unions in Britain, and which have been their mainstay throughout the period of their formation and development to maturity. It is the union membership in the mining, railways and shipbuilding industries that has been the backbone of the movement, and the massive reduction in their membership and consequently in their influence is bound to have an effect on the future character of trade union activities.

These are the industries where trade union consciousness has been the strongest, where the class struggle has been fought with a ferocity not experienced in the newer industries and services, where the trade unions have borne the brunt of successive attacks by employers and suffered the social misery of heavy unemployment and distress down through the years. These are the unions which have provided the colourful history now recalled in the Centenary Brochure of the TUC. Not only have they been the strongest in trade union consciousness, they have been the strongest, too, in political consciousness. Their communities are strongholds for labour politics, and labour politics with a dynamic socialist content.

The qualities they have given to the trade unions will diminish to be substituted by the qualities of the membership which will come from sectors where growth is taking place. The leading role of these unions in the inter-war years, within industry and the Trades Union Congress, exerted the decisive influence that determined the policies adopted and prosecuted by the movement for most of its existence. Because of the changes that have taken place and which will continue, this leading role will disappear to be taken over by unions operating in expanding sectors of the economy and by the importance of their membership within the economy.

The process of growth in the working population is not to any significant extent being reflected in a growth of union membership. Apart from some more recent new affiliations of organizations, membership remains static when it obviously ought to be growing if only to keep pace with the growth in working population. In this context it is declining, with TUC membership representing a smaller proportion of the total working population. It is, understandably, difficult to measure union growth by industry or occupation, since most unions do not follow any system of classifying membership by occupation, and in any case there will be non-manual grades included in the membership of most unions. The potential for union recruitment is greater in this field than in any other, and if recruitment results in substantial increases in membership in the years ahead, as it certainly should, the character of trade unions in Britain will undergo radical change.

The ratios between manual and non-manual workers within the movement will change with the result that the influence of the non-manual section will increase, and this increased influence will find greater expression in the determination of trade union policy. If the organization of non-manual grades is to take the form of horizontal unions then the number of unions operating in industry could increase, or their strength increase with the possibility of separate bargaining procedures adding to those already in existence. It will be in the best interests of unity within the trade union movement if the non-manual grades and occupations are, where possible, incorporated into industrial unions.

The unions in the contracting industries brought into the trade union movement an enlightened and coherent concept of unionism which had been nurtured in their past struggles and which became part of the tradition and heritage handed down. Part of this heritage was the principle of solidarity, of acting collectively, of thinking and acting not on the basis of individual interests but on that of the mass. This unity was the product of a century of conflicts in which the older unions have participated.

It is true, of course, that those who fill the non-manual occupations now and in the future will be the sons and daughters of parents who are products of the older and heavier industries. But the tradition and spirit of an industry has little force outside the place of its origin; it is not a quality that is transferable. The point of this argument is that the type of union member to be recruited from the growth sector will almost certainly be a different kind of social being to those now comprising the main body of union membership. Background, social strata, educational standards, social and cultural interests, these will be the new factors influencing the character of this new force entering the trade unions.

Although the effect of common grievances and interests arising from common employment will tend to unite and produce a collective approach to these common problems, there will be more emphasis on the individual interests. In the area of non-manual occupations the concentration of work-people within the place of employment is appreciably smaller, and the force which was represented by industries that were, and still are, labour intensive, will be largely non-existent in this sector. Consequently there is not likely to be the same intensity and cohesion of agreement and action.

This is not being derogatory to non-manual occupations, but it is an attempt at an appraisal of the effect that this kind of change in the composition of the membership of unions can have upon policy and activities. It is based, too, upon an experience of working with and for such employees in the mining industry.

The difference I have found is that people employed in these grades who by the nature of their jobs have to operate in the main as individuals, demonstrate a stronger element of individualism within the trade union. They are more critical, with a greater capacity for critical analysis, are less influenced by oratory and rhetoric and more reliant on their own individual logic and reason. These are good qualities making for better discussion and understanding, and reducing the extent of support for policies and actions based on emotionalism or blind loyalty.

The effect of this new quality upon strategy and tactics is that union leaders will need to think ahead and plan for the future, strategy will need to be carefully thought out and presented in detail and adequate provision made for participation by these members in the formulation of policy. This will be an occupational section who will not display the same degree of willingness to 'leave it to the leaders' and who will not be very concerned with tradition and precedent. It is true that there are already a large number of 'white collar' members in unions, but the future must be to branch out into new spheres to bring many more into membership, with the inevitable consequence that these occupations may in future years replace in power and influence the older unions now in the throes of contraction.

The second important factor to be considered is, that in the prosecution of their historic purpose in this changing environment, trade unions will need to be concerned with adapting their programme to meet the new conditions. This, of course, is an aspect of function that needs to be constantly under review. This was recognized back in 1944 when the TUC outlined a perspective and programme for the post-war years. The programme projected then was to push ahead with improving the living standards of the people, to preserve full employment and to extend workers participation in the control of industry. The more specific resolutions carried at successive annual congresses provide a miscellaneous addition to this more general statement of strategic aims.

The dynamic of the trade union movement and its source of inspiration and attraction for working people have always been the aims it has set itself which have crystallized their aspirations and needs. The high points of popular support in union history have been associated with its campaigns for urgent reforms. The highlights of greatest support and interest in the mining industry were around demands like the eight-hour day, for minimum wage, for paid holidays, sickness pay. In the main, these are reforms which have now been secured, although they need constant improvement to meet modern standards, and demands for such improvements form a regular part of annual Conference Agendas although they no longer inspire great interest among members.

Reforms of this kind are no longer to be within the capacity of individual unions as was the case in the past. They will need to be the concern of the Trade Union Congress involving as they now do discussions with central Government, and Government agreement before they can be realized. The responsibility for stimulating interest and support for major reform is now with the TUC. The production by them at the beginning of each year of an Economic Review is an excellent development and its proposals to stimulate economic growth, for maintaining full employment, and norms for wages increase are very necessary and important for facilitating its role in the planning counsels of the nation. From the standpoint of inspiring the confidence of the working population or the members of the trade union movement, this review is of little consequence.

What is required is a programme of longer-term aims which meet the needs not only of union members but also of the working people generally. The presentation before the members and public of such a programme is vital in order to project the TUC as a dynamic centre of the movement with enlarged responsibilities, and to create an interest in unionism among the vast mass of non-members making up the bulk of the labour force. It is not enough to say that the long term aim is socialism and that this is our inspiration. This for some of us may be true but it is unlikely to cause any great stirring in the breast of the worker who faces job insecurity, or redeployment, or any one of the many upsets he is told is caused by the new industrial revolution which appears to him more as a menace than a benefit.

What the TUC needs now is a charter of long term worthwhile aims, related to modern conditions and perspectives designed to meet the needs and problems which arise in the working and social lives of working people.

A start has been made with the adoption of the aim of a £15 minimum wage, although this is much more a short term rather than long term aim. In addition, the demand for an extension of industrial democracy, and for improvements in reforms already operating, represent important claims, but these are so vaguely stated, so indifferently pursued and so indefinite in relation to timing of achievements that they are seen as pious expressions of hope rather than a charter for action. What is needed is something more than a statement of general purpose although the statement will include aims of great interest to members, such as the maintenance of full employment; compensation for loss of employment not less than average earnings; protection of the real value of wages; the extension and adaptation of social and cultural amenities to meet the new work schedules and many other useful reforms which are close to the needs of people but excite little enthusiasm.

Would it not arouse interest and support if a Charter of Aims was devised to be achieved in five years to be campaigned for by the TUC and the affiliated unions? If the gap between organized and unorganized workers is to be closed it will need colourful agitation around a series of aims which will capture the imagination of the activists in the unions, the rank and file and the public. In addition to the aims already referred to I would suggest the following Charter of Reforms to be achieved by 1975.

(1) A minimum wage of £20 based on existing values.
(2) One month paid annual holiday with State-run schemes for holidays abroad.
(3) A 35 hour working week spread over five days.
(4) Equal pay for equal work irrespective of age or sex.
(5) The provision of guaranteed wages based on average earnings if, for reasons over which the worker has no control, work is unavailable.

The above by no means exhaust the reforms that are required and which could be included in the Charter. They represent the kind of aims that I believe are needed to arouse and restore the kind of interest in the trade union movement which characterized its earlier history.

IS THERE A DANGER OF A CORPORATE STATE?

There is no progress without risk and danger, and there could be definite dangers arising from the trends in the functioning of the trade union movement in present conditions. The increasing role and power at the centre could increase this danger. The involvement of the trade union movement with employers and Government over the whole range of their joint activities could be at the risk of crystallizing into an eventual unity of purpose. Indeed, some trade union and political leaders already accept this as being proper and in the interests of union members. The view of such leaders is that the interests of workers, employers and Government are are identical on major questions, and that this concept of a social relationship should replace that of a movement of workers struggling for the progressive change and ultimate overthrow of the present social system. This outlook, of course, is embraced by those who have never recognized the division of society into classes or the existence of a class struggle.

The danger is that the forms of joint activity can consolidate into a permanent corporate unity in which the independence of the unions and their special interests may be lost in the development of a more closely knit joint organization. The ultimate in this form of organization was the corporate states of fascist Italy and Germany, and the present set-up in Spain, but there are, of course, many possible permutations of the systems which operated in these countries. The big problem for the trade union movement is to be able to keep a balance between joint involvement with the Government and employers considered necessary to the pursuit of union aims, and a degree of involvement that submerges its own identity into an integrated corporate organization.

The danger of this kind of philosophy assuming some domination is at its highest when a Labour government is in power. The ties between a Labour government and the trade unions is, by their history, of the closest kind, and the personal relationships between the leaders of both institutions are strong and intimate. Although there can be sharp conflicts between them creating temporary strains, this basis of close mutual relationship is sustained.

The danger in such circumstances is that of the trade union movement becoming an aide to the Government, and because it is a friendly government, accepting measures that would be strongly opposed if put forward by a government considered unfriendly. The TUC in the inter-war years has been described as being 'on the outside looking in'. It has now achieved a status where it is on the 'inside', but the 'inside' can be as understood by the old lag. It is possible for the inside to become a voluntary prison.

The danger is not so much one of organizational unity with employers and State, but of a more insidious unity expressed as a political philosophy which in practice could put trade unions in the position of adapting their policy to that required by the government and employers. In recent months the TUC has appeared to do this, although in truth the accommodation of Government policy was unwillingly adopted. This could be said to have been the position when the TUC decided to operate its own incomes policy. The design of the tactic was to obviate the necessity for the Government's legally enforcable Incomes Policy. The same comment is possible on the measures to deal with unofficial strikes. Thus, objectively, the trade union movement is put into a position of appearing to accommodate features of Government policy at the instigation of the Government. Whatever may be said of the arguments and recriminations surrounding these policy innovations, they are inspired by policy requirements of the Government.

The point being made is not whether the decisions to accommodate Government policy have been right or wrong but that they represent what could develop into a trend where the trade union movement is more and more adapting its policy to meet that of the Government. In certain political conditions the end product of such a trend could be corporatism and not socialism.

If the measures adopted by the TUC are justified and necessary, and in existing circumstances they are, then their introduction should have been based on the independent assessment of the situation by the TUC itself, and not from a judgement made by the Government.

To avoid situations of this kind and the long term dangers they represent, it is not enough to shout slogans for the 'freedom and independence of trade unions'. There is no argument that if a drift towards a corporate state situation is to be avoided then the freedom and independence of the trade union movement must be preserved and strengthened. But this can only be done by unions facing up to the realities of this changed situation in which they now function. The reality is that much of what was treasured in the past, in forms of organization, the methods of wage bargaining, the wages systems, and the concepts of freedom and independence which were restricted to sections and individual organizations, are outmoded and have no place in the outlook and operations of trade unions in modern conditions.

The operations are no longer concerning a local boss, or in most situations not even a national boss, but are concerned with powerful national and international organizations, with the State being involved in any and every major demand for industrial and social reform. In fact, in the years ahead, international trade union organizations to which British unions are affiliated, will be called upon to play a part as international federations not unlike the part played by the federation of unions in the countries that make up the international.

As this study has sought to emphasize, this is a new situation and the trade union movement, if it is to function effectively, must be involved deeply in the machinery for determining national policy, and this in turn necessitates changes in trade union strategy and organization. Equally, the concepts of freedom and independence of unions appropriate to a past that is gone, must be adapted to the present that is here and the future ahead.

SOME FINAL COMMENTS ON TACTICS

There always has been and probably always will be argument and controversy on the question of what in any given situation are the best tactics for unions to operate to attain a given end. It is perhaps of some importance to examine some of the arguments being advanced at the present time on this theme. At some time or other in a trade union lifetime one has been involved in most of the arguments and the tactics they range around. There is no book of 'recipes' from which to select to meet a particular situation. The general context of the argument for the tactic based on 'action' is roughly along the following lines.

'Workers and the unions which represent them exist and function in a capitalist society which exploits them and their struggles are therefore against exploitation. The fight for increased wages and other improvements is part of the general struggle for economic and social justice. The capitalist system is a 'free for all' system using any means to get for those who own property in one form or another, the maximum 'rent interest and profit', in a competitive economy. The trade unions are therefore justified in using any and every means available in the prosecution of their aim for justice and freedom'.

The definition of capitalism and its operations fits in with that accepted by most socialists and is thus acceptable, but some reservations must be stated against 'free for all' tactics by workers and unions.

The first reservation, of course, is that no one wants to promote or engage in struggle for its own sake, although statements by some people on occasion give this impression. But no responsible political or trade union leader would advocate such tactics, and if they did they could not last long in credit with workers to whom they gave such advice. There must always be a good reason for action, where the purpose involved justifies the means being used.

There are many meanings possible to the terms 'action, struggle, fight', but usually they are meant to have a connotation with 'strike'. But whatever the term or the definition given to it, struggle or strike action is not an end in itself but a means to an end. The end must justify the means, and not all strikes satisfy this test. Strikes are not simple events but can be very complex once they start, creating problems of mass psychology and moods that are often more difficult to resolve than the issue that caused the action. On many occasions I have addressed mass meetings of miners while on strike and have been in a position to state that, immediately there was a resumption of work a meeting with the employers would take place with the prospect of their grievance being settled satisfactorily. More often than not, talks took place with the employers while the strike was in progress and the demands of the men conceded, but on the strict understanding that they were not to be made known until there was a resumption of work. It was generally accepted in the mining industry that it was not in the interests of either the union or the employers to announce settlements while strikes were on, because the incidence of strikes were so frequent that such an inducement would have made things worse. But often the pledge of a guaranteed settlement would not secure a return to work. Sometimes there could be a build up of grievances over a period of time which resulted in a sudden explosive outburst, elemental in its fury. The particular issue which fused the explosion could be settled quickly but the force and heat produced could take some time to taper out. Reason and persuasion carry little weight in situations of this kind.

One must have reservations, too, in respect of lightning sectional strikes, which take place without any prior notice or consultation with the union or other workers who may be immediately affected. It is true, of course, that situations arise suddenly where men are not prepared to accept or aquiesce in a condition even for the short time it might take for the normal procedures of conciliation to be brought into operation. But it ought to be possible to reduce the incidence of this kind of action by close consultation between union and management at an undertaking, since the issues which lead to the action are usually capable of being anticipated and avoided. Very often, too, these actions can be frivolous and over questions that are already being negotiated, or where a settlement could reasonably be expected by negotiation.

The justification for lightning strikes should involve some vital principle, and be something more than the run of mine or mill grievance. Action by a section without consultation with the general body of work-people who may be put out of work, is a position where the 'tail is wagging the dog' and sooner or later leads to internal strife. From my own experience I have had to face bitter fights between union men, in circumstances where the men at a coal-preparation plant serving a number of pits have come out on strike, and the pits made idle – a few hundred men putting thousands temporarily out of work over a dispute they knew nothing about and the settlement of which would not provide any benefit to them.

The principle of solidarity between workers is an admirable quality, but exploiting it in the way that is sometimes done can lead to its destruction. The practice of acting first and talking after, may be all right as a relation between enemies but it can hardly be extolled as a satisfactory relationship between friends.

I have personal reservations, too, on the development of unofficial and unconstitutional organization and leadership within trade unions. The aim of such an organization is usually to organize action independent of the union and often in direct opposition to union policy. It may be that the internal position in some unions lends itself to this development, as well as the isolation and remoteness of leadership. This isolation often leads to inattention to members' problems, and union policy and outlook far removed from that supported by the members. There is little doubt, too, that the multiplicity of unions in certain industries gives impetus to leadership on the spot, as well as the separation of shop stewards from formal union organization. These are factors that obviously represent fertile soil for this type of growth.

In the South Wales Coalfield, even with a militant leadership and an executive committee comprised of men employed in the pits, this unconstitutional movement came into existence. For a time and on matters arising in the industry, unofficial conferences were called, sometimes involving half the coalfield, a leading committee was elected to deal with matters arising between conferences; in general an organization within an organization.

The position had to be fought and overcome, not on the basis of taking disciplinary measures or suspensions from membership, but by argument and discussion and eventual understanding that such divisions weakened the union in relation to the employer and if persisted in could lead to the union's collapse.

Very often the stimulus to this form of organization even in unions regarded as progressive, is the slowness of the official machinery in resolving problems referred to it. On the other side of this there is very often an unreasonable impatience which stems from a failure to appreciate the difficulties involved in a claim, and this is usually due to a failure of communications within the union. But duality of leadership must cause divisions, confusion and internal strife. Very often it leads to attempts to impose discipline by union leadership and this tends to aggravate and intensify the antagonisms already existing. Without wishing to pontificate further on this form of unofficial movement and dual leadership except to state that, from my own experience, while it might show temporary results initially, the disunity it creates can have long-standing disabling effects.

The philosophy of struggle against capitalism is progressive and a necessary condition to the advance of socialism. This means surely that the forms of struggle must not only be directed to weakening capitalism, but must at the same time strengthen the understanding and movement towards socialism.

No one knows better than I do how difficult it is when men are incensed by an attack on their rights or conditions of employment, how futile it is at that moment to argue for restraint on the grounds that the action about to be taken will not strengthen the cause of socialism. As stated earlier, situations arise when only prompt action by the men in industry can secure immediate redress. But just as strikes are not ends in themselves, not all strikes that have 'ends' are necessarily justified or progressive. They can very often be for very selfish and sectarian ends and contribute nothing to social advance. To make a fetish of struggle or strikes as the only means of securing remedies to grievances is as stupid as the outlook that it should never be resorted to.

What is important, if there is genuine concern for developing trade union and political consciousness out of the struggle for improvements, is that such concern must also include the effect of an action on other workers and people who may be directly affected. This must always be the concern of those considering strike action. Obviously a strike would be futile if it did not dislocate and disrupt and thereby exert pressure on those against whom it was directed. But there must be consultation and understanding of the reasons for the action if solidarity and support is expected.

The right of workers to strike is fundamental and must be preserved. But it is a weapon that in modern conditions has a far more extensive chain of consequences upon other workers than in years past. For this reason its use should not be abused and applied indiscriminately. As stated earlier a major strike in the present economic conditions, because of the chain system of production and the interdependence and interrelation of industries, can have nation-wide effects. A dispute that affects a very small number of men can affect the employment of thousands who are remote and unconnected with the issue or the settlement. In fact, as distinct from the inter-war years, a sectional strike in a key industry can be almost as extensive in its effects as was the General Strike in 1926. This is not a case against the use of the strike weapon, but it is an argument against its irresponsible use.

The ramifications of the big production enterprises are international and the dislocation of production in one country may hamper and restrict its rate of profit, but their vulnerability is small, and their flexibility to shift the emphasis of their activities gives them substantial protection against total dislocation. It is this ability to frustrate effects, plus the effect the action in these circumstances can have on masses of other workers which creates doubts in my mind as to the long term effects upon the solidarity so vital to the trade union movement. These are being accepted, as I saw them accepted in the South Wales Coalfield, but the long term effect was a violent reaction which led to the 'tit for tat' strikes already referred to. Trade union and political consciousness is not advanced in the climate produced by indiscriminate strikes that have such far-reaching effects.

There is of course a budding outlook that the development of struggle on any and every possible issue for which support can be got, will lead to the collapse of capitalism. That in the process of this collapse the workers and people will become revolutionary, creating their own organizations and leadership, and that out of the anarchy created the answer to all social problems will emerge to provide the basis for a new social order. All my experience contradicts this acclaimed revolutionary doctrine. In the early 1930s there was the collapse of government in this country, the police, army and navy sweating under cuts in their wages, mass misery and poverty, but no revolutionary spirit. Agitation, campaigning there was by those of us who were revolutionaries, but the response for years was feeble and fragmentary. It does not follow that chaos in economic and social conditions creates the conditions for revolutionary change.

It is the response of people to situations of breakdown, and the constant agitation and organization on issues arising from their daily lives, and keeping in the forefront the concept of a new social system based on social justice that represents the constant hard grind to produce social change. There is no easy spontaneous way.

The foregoing, I anticipate will be branded as a thesis against strike action. This, however, is not the intention, since the abandonment of this right either voluntarily or by the force of legislation takes away the source of trade union power. In fact the attempt to take it away by legislation would create the opposite effect to that which the legislation would be intended to achieve. For the trade unions to abandon it voluntarily would lead to a flood of unofficial actions far beyond the incidence being complained about now.

These comments attempt to stress the changed environment in which they now take place, and the far greater range of their effects. From the standpoint of those who do not care about effects, this of course adds to the power of the action. But the power can be destructive when set against the interests of the working class as a whole. Struggle there must be but it can have many forms which are effective in exerting pressures and moulding opinions for progressive change. The concentration needs to be on forms that unite and build the power of the trade union movement.

CONCLUSION

Those who have read this study will recognize that it has no merit as an academic or literary exercise.

It is no more than a sketchy presentation of contrasts, questions and conclusions, laced fairly liberally with personal reminiscences. It ends as it started, with the hope that it will provoke some discussion within the trade union movement on the questions and assessments which have been put forward.

The purpose of the study has been to examine the changes that have taken place in the role of the trade unions. In the last half century the changes in the economic and social environment have been dramatic and fundamental. The contrast between the function, status and authority of the trade union movement in the inter-war years and in the present is a measure of the progress it has made.

Although this progress has been great, the opinion stressed in these pages is that it has been inadequate to meet the challenge now and in the future of the new power centres which have been created. The main considerations upon which this charge of inadequacy is based can be briefly enumerated to help focus discussion.

First, central Government involvement directly in industrial relations, prices and incomes policy is not temporary, but will be permanent and is likely to increase. The main reasons for this are the financial and economic interdependence of nations, the influence of world banking, balance of payment problems, the cost of the technological revolution and its effects, and the fact that this increasing role of Government is a feature of all capitalist countries.

Second, the rapid growth in all sectors of the economy of huge international trusts wielding massive economic and political power and operating in several countries, thus reducing their vulnerability to pressure in any one country.

Third, the effects of the technological revolution, based on automation and remote control systems, on the content of jobs and job relationships and, because of this, on wages systems and on inter-union relations in multi-union industries.

The combination of these factors represents the main source of challenge to the trade union movement, compelling a judgment as to whether its present organization, strategy and power distribution can effectively and successfully meet the challenge.

The conclusion of this study is that there is no basis for confidence that it can, and the reason for this lack of confidence is basically that in face of this centralization of power on the side of the international trusts and central Government, the power of the trade unions continues to be dispersed and fragmented.

It is part of this conclusion that unions are clinging to forms of organization that dissipate their strength and power. As with the employers and Government, their power should be concentrated in each industry and centralized nationally, on questions involving political action, in the Trades Union Congress.

The suggestions put forward to overcome these shortcomings are of themselves a challenge to the more hidebound concepts of union structure and purpose, which are now holding back effective reorganization.

The policy of mergers and amalgamations does not solve the central structural problem, that of multiplicity and overlapping of unions and it can have the effect of consolidating and thereby accentuating it.

In industry the problems of wages systems and wage relativities resulting from changed methods of operation will be continuous, and without single-channel bargaining procedures can lead to inter-union conflict.

Major reforms affecting hours, fringe benefits and all problems of general industrial policy will not be achieved by negotiation between unions and employers, but can only be brought about by discussion between the trade union movement as a whole and the central Government.

Fragmented trade union action, in reaction to the power and range of the new 'conglomerates', will be less effective, with a greater risk of disaffection within unions due to the chain effects.

Two main conclusions follow. Firstly, there is urgent need to restructure the trade unions towards vertical industrial unions and away from the horizontal, multi-union structure of the present. It is not only desirable to have one union facing one management or one company, but one union facing the employers' organization in one industry. Secondly, equally important is the conclusion that individual unions must be prepared to subordinate their own autonomy and authority to the Trades Union Congress and its General Council in respect of policies and functions where the central Government and the Confederation of British Industries are concerned.

These adjustments may be hard to make, but the failure to make them could be disastrous for the Trade Unions movement.